Contents

Manufactured Housing:
Regulation,
Design Innovations,
and
Development Options

Welford Sanders

P a r t 1

Current Trends in Manufactured Housing Regulation

Manufactured housing is fast becoming the only housing that a growing number of Americans can afford to own. In recent years, well over 300,000 manufactured homes have been sold annually, and, according to the U.S. Census Bureau, about one-fourth of all single-family housing starts in 1996 were manufactured homes.

n 1986, the American Planning Association (APA) published *Regulating Manufactured Housing* (Planning Advisory Service Report No. 398). Based on a survey of the more than 1,000 communities that subscribed to APA's Planning Advisory Service at that time, the report revealed that both the quality and appearance of manufactured housing had improved significantly over the previous 10 years. In fact, some manufactured homes were found to be virtually indistinguishable from site-built homes.

The construction of these factory-built units is regulated by the U.S. Department of Housing and Urban Development (HUD) under the Manufactured Home Construction and Safety Standards (HUD Code). This national building code was enacted in 1976 to improve the quality and safety of homes that were at that time referred to as "mobile homes." In 1980, the official designation of these homes was changed to "manufactured homes," both in recognition of their improved durability and less mobile nature—once they are sited, manufactured units are rarely moved.

The survey analyzed in the 1986 report also found that more communities than ever before were allowing manufactured homes in single-family districts. It also found that the general public's acceptance of this factory-built housing was at an all-time high.

Since the publication of the 1986 report, both the quality and appearance of the manufactured home has continued to improve. A 1993 PAS Report, the *Manufactured Housing Site Development Guide* (PAS Report No. 445), examined some of the innovations in unit and manufactured housing development designs at that time.

But the single most important attribute of these factory-built homes remains their affordability. The latest data from the U.S. Census Bureau indicates that the per-square-foot retail cost of a site-built home in 1993 was twice as much as the $25.97 per-square-foot cost of a multisection manufactured unit. Manufactured housing is fast becoming the only housing that a growing number of Americans can afford to own. In recent years, well over 300,000 manufactured homes have been sold annually, and, according to the U.S. Census Bureau, about one-fourth of all single-family housing starts in 1996 were manufactured homes.

Today's manufactured homes are bigger—in 1996 the number of larger multisection manufactured units purchased annually surpassed the number of smaller single-section homes being sold. For the first time, two-story manufactured units are now available. Increasingly, manufactured homes are being used in subdivision development. Furthermore, a growing number of developers and builders who have not worked with manufactured housing before are now involved in manufactured housing development. Manufactured housing is currently being designed for urban infill development. In a number of cities, manufactured units are being sited in inner-city neighborhoods as an alternative to more costly site-built housing.

More communities are permitting manufactured homes in their single-family neighborhoods and in many cases as a use by right. Communities also appear to be taking a different attitude toward manufactured home land-lease or rental developments and now view them correctly as residential uses that belong in residential zoning districts. A number of local governments have established a single set of zoning provisions for all types of single-family housing, applying the same standards to site-built and manufactured housing. A number of states require that local governments take steps to ensure that their regulations do not discriminate against manufactured housing. Moreover, the courts in some states have ruled that local governments may not impose restrictions on manufactured housing without imposing those same restrictions on site-built housing.

While the regulatory climate for manufactured housing has improved since 1986, there are still many communities that continue to carry over outmoded stereotypes and restrict, often unnecessarily, the use of manufactured homes. In some cases, these communities are not aware of the new and improved manufactured units that are now being built, nor are they aware of how a growing number of local governments have instituted regulatory provisions that allow for and encourage well-designed manufactured homes and developments in the same zoning districts as site-built housing.

This report examines current local government regulation of manufactured housing. It should be especially useful to those communities that want to update their zoning provisions for manufactured housing. This report explores a range of regulatory approaches and development standards so that local planners can choose the techniques and controls that are best suited to their circumstances. Changing public attitudes toward manufactured housing are examined, along with current trends in the design and construction of manufactured homes.

The specific focus of this report is on zoning provisions that allow manufactured housing to be regulated as a residential use in residential neighborhoods. The scope of this report is limited, for the most part, to zoning ordinance treatment and subdivision control of manufactured housing. Modular and other types of factory-built units that are built to state or local building codes are not examined, although they also represent a growing share of the housing market. These housing units continue to enjoy greater community acceptance than manufactured housing and are almost always treated like site-built homes in zoning regulations. These units are typically more costly to build than manufactured housing that is built to a national performance-based code and are, therefore, less affordable. Building codes or construction standards are also beyond the scope of this report. A brief discussion of unit installation requirements, however, is provided in Part 2.

THE 1996 SURVEY

This report, like the 1986 report, is based on an APA survey of communities that subscribe to APA's Planning Advisory Service. The current survey of 1,172 communities was conducted in the summer of 1996. It sought information about local zoning and subdivision regulations governing manufactured housing. It included questions about how communities distinguish between the different types of units built in a factory; where manufactured homes are permitted in the community; and under what circumstances this housing may be allowed. The survey also asked about the general attitude toward manufactured housing and if local officials had experienced any problems related to the design, durability, or other aspects of this type of housing. The communities were also asked to send APA their zoning provisions for manufactured housing and related local studies or reports and information about manufactured home developments in their community that could serve as models for other communities.

There were 508 responses to the survey, which is a response rate of 43 percent. This report is based on the 475 returned questionnaires that were found to be complete and "usable." Other information used to prepare this report was gained from visits to Southern California; Baltimore, Maryland; Wilkinsburg, Pennsylvania; Denver and several other Colorado communities; and North Carolina.

Table 1 presents some key findings of the current survey and two earlier surveys: a 1970 survey of PAS subscribers reported in *Modular Housing, Including Mobile Homes: A Survey of Regulatory Practices and Planners' Opinions* by Frederick H. Bair (PAS Report No. 265, January 1971) and the 1985 survey noted above.

The results of these surveys indicate a trend toward greater acceptance of manufactured housing and fewer restrictions being placed on this type of housing over the past two decades. The 1996 survey found that nearly 90 percent of the respondents indicated that they now permit manufactured homes in residential districts. Similarly, 88 percent of the communities that permit manufactured units in residential districts allow such homes outside of manufactured housing developments and on individual lots.

The 1996 survey found that nearly 90 percent of the respondents indicated that they now permit manufactured homes in residential districts. Similarly, 88 percent of the communities that permit manufactured units in residential districts allow such homes outside of manufactured housing developments and on individual lots.

Table 1. Regulatory Treatment of Manufactured Housing: 1970, 1985, and 1996 Survey Results			
	1970	**1985**	**1996**
Number of communities responding	NA	359	508
Usable returns	NA	234	475
Communities that permit manufactured homes in residential districts	NA	70%	88%
Communities that permit manufactured homes on individual lots	38%	60%	88%
Permitted by right	1%	52%	83%
in most restrictive residential district	NA	NA	59%
in all residential districts	NA	NA	53%
Parity in the regulation of site-built and manufactured housing	NA	NA	29%

The increase in the percentage of communities that permit manufactured housing by right was not as great in the last 10 years—an increase of 32%—as it was from 1970 to 1985 when the percentage of communities permitting manufactured housing by right increased from 1 percent in 1970 to 52 percent in 1985. Nevertheless, the increase that did occur between 1985 and 1996 was substantial.

Further evidence of the trend toward less restrictive zoning treatment of manufactured housing was provided by responses to three questions in the 1996 survey that were not asked in the earlier surveys (see the last three items in Table 1). These questions have to do with the zoning districts where manufactured housing is permitted and whether communities apply the same regulations to manufactured housing and site-built housing. The current survey found that nearly 60 percent of the communities responding indicated that they permit manufactured homes in their most restrictive single-family residential districts, and over half of the communities surveyed reported that they allow manufactured homes in all residential districts.

There were two categories of responses to the question, Does your zoning ordinance include specific regulatory provisions for manufactured housing? In most cases, when the response to this question was No, it meant that the community's zoning ordinance did not include provisions for manufactured housing and that this type of housing is regulated by another ordinance (in some cases, the building code) or jurisdiction. This category of respondents usually did not complete the survey since most of the remaining questions would not be applicable. Nearly a third of the communities that answered No to the above question explained that this did not mean that their zoning ordinance did not include regulatory provisions for manufactured housing; rather, it meant that their zoning ordinance did not include specific regulatory provisions for manufactured housing that did not also apply to site-built housing. In effect, these communities treated manufactured units and site built units equally. The following response of one local planner typifies the approach taken by those communities who regulate manufactured housing and site-built housing equally:

> [Manufactured housing] . . . is regulated in the same manner as any other type of single-family dwelling . . . is allowed in all five of the city's residential zoning districts. . . . Manufactured homes are subject to the same zoning regulations as any other type of single-family dwelling.

This community and others view the manufactured home as a dwelling unit and include manufactured housing as well as site-built housing in the community's legal definition of a dwelling unit. In such communities, when there is concern over whether a certain aspect of factory-built housing is compatible with site-built housing, there are usually standards applied to both factory-built and site-built homes that attempt to ensure such compatibility. A common standard for both types of housing, for example, is a minimum unit-width requirement. This and other development standards for manufactured homes are examined in greater detail in Parts 2 and 3 of this report.

PUBLIC ACCEPTANCE

The 1996 survey found that a third of the local officials (32 percent) responding to the 1996 survey reported that the general public attitude toward manufactured housing was positive, compared to 40 percent reporting similar sentiments in 1985. (See Table 2.) Planners in communities where there was resistance to manufactured housing said that such attitudes are due primarily to the continued existence of old mobile home or trailer parks and/or the perception that manufactured homes have an adverse impact of property values. One planner notes that "We have some very poorly managed and maintained older mobile home parks that have given all of this housing a poor reputation." This same planner, however, reported that, in recent years, the general public perception of manufactured housing has moved toward greater acceptance, noting that "the industry has worked

The current survey found that nearly 60 percent of the communities responding indicated that they permit manufactured homes in their most restrictive single-family residential districts, and over half of the communities surveyed reported that they allow manufactured homes in all residential districts.

6

	1970	1985	1996
Table 2. Public Acceptance of Manufactured Housing: 1970, 1985, and 1996 Survey Results			
Communities reporting a positive public perception of manufactured housing	25%	40%	32% Most common reasons given for unfavorable attitudes toward manufactured housing were the presence of old parks and the perception that manufactured housing has an adverse impact on property values.
Changing perception of manufactured housing		Mostly positive and toward greater acceptance	79% of the communities surveyed indicated that the general public's perception of manufactured housing had changed, and this change was toward greater acceptance of manufactured housing.

very hard to improve standards." This community was one of a number of communities which reported that, even if the general public perception is still clouded by the view that manufactured housing means mobile homes, there is greater acceptance of manufactured housing. In fact, 79 percent of the 508 communities which indicated that the public perception of manufactured homes has changed over the past few years further indicated that such change has been toward acceptance of manufactured housing.

Some communities and a growing number of state associations of the Manufactured Housing Institute (MHI) are taking steps to address the problems associated with old mobile home developments. Some local governments, for example, now encourage or require that these developments be improved. Some examples of how this is being done are examined in Part 2. Similarly, manufactured housing industry state associations, with support from MHI, are sponsoring workshops that show industry personnel how the maintenance and overall appearance of old developments may be enhanced.

The industry has also supported research into whether manufactured housing has an adverse impact on the value of existing site-built housing. Over the past decade or so, several studies have examined this issue. These studies all reached the same conclusion: manufactured housing and manufactured housing developments have not been shown to have a negative effect on either the market value or appreciation rate of adjacent site-built housing.

One recent study, for example, conducted by the University of Michigan in 1993, considered whether an existing manufactured housing development had any effect on site-built subdivisions developed after the manufactured home development was in place. It also studied whether a manufactured housing development built after the subdivision was completed had any

Some communities and a growing number of state associations of the Manufactured Housing Institute (MHI) are taking steps to address the problems associated with old mobile home developments. Some local governments, for example, now encourage or require that these developments be improved.

Manufactured housing and manufactured housing developments have not been shown to have a negative effect on either the market value or appreciation rate of adjacent site-built housing.

effect on the subdivision. The researchers concluded that "rental manufac-tured home communities did not appear to have a significant effect, positive or negative, on adjacent residential property values" (Warner and Johnson 1993, i).

A 1997 study conducted by East Carolina University (Shen and Stephenson 1997) of manufactured housing developments in several North Carolina counties found "no clear negative correlation between the overall apprecia-tion rate of site-built residential properties and the presence of manufac-tured housing in close proximity."

REGULATORY PROBLEMS AND ISSUES

The regulatory "negligence" and abuses that were common two decades ago (e.g., restricting manufactured homes to inappropriate, nonresiden-tial districts) are far less common today. The 1985 survey, for example, found that 11 percent of the responding communities did not, under any circumstances, permit manufactured homes in their jurisdiction, and more than a quarter of the communities that did allow manufactured housing indicated that these homes were not permitted to locate in residential districts. The 1996 survey found that only about 2 percent of the communities responding do not permit manufactured homes in their jurisdictions, and only about 13 percent did not allow manufactured housing in residential districts.

The 1996 survey made it clear that there is one area where there has been little change in the treatment of manufactured housing, and that concerns the terminology used to describe or identify manufactured housing. Many communities continue to refer to manufactured housing—the official des-ignation of units built to the HUD Code—as mobile homes.

STATE STATUTES AND THE COURTS

In 1985, only 14 states had passed legislation that prohibited exclusionary and unfair regulatory treatment of manufactured housing. By 1996, 22 states had such legislation. (See Table 3.) More states than ever before require that local government treat manufactured housing and site-built housing equally in their regulations. As was the case in 1985, most of the communities that permit manufactured homes in their more restrictive residential zoning districts and as a use by right are located in states that enacted such legislation.

The courts in at least four states where antiexclusionary provisions have not been enacted have ruled that local governments must accept manufac-tured housing. Some states promote the use of manufactured housing locally as part of their broader efforts to encourage local governments to provide more affordable housing. (See Table 3.) The General Assembly of North Carolina, for example, has put forth the following directive to local governments:

> The General Assembly finds and declares that manufactured hous-ing offers affordable housing opportunities for low- and moderate-income residents of this State who could not otherwise afford to own their own home . . . [and] further finds that some local governments have adopted zoning regulations which severely restrict the place-ment of manufactured homes. It is the intent . . . in enacting this section that cities reexamine their land-use practices to assure com-pliance with applicable statutes and case law and consider allocat-ing more residential land area for manufactured homes based upon local housing needs. (Section 160A-383.1 Zoning Regulations Manu-factured Homes)

The 1996 survey made it clear that there is one area where there has been little change in the treatment of manufactured housing, and that concerns the terminology used to describe or identify manufactured housing. Many communities continue to refer to manufactured housing—the official designation of units built to the HUD Code—as mobile homes.

Most of the communities that permit manufactured homes in their more restrictive residential zoning districts and as a use by right are located in states that enacted legislation requiring equal treatment of manufactured and site-built housing.

Table 3. State Statutes and Court Decisions Related to Manufactured Housing

Passed Legislation that Prohibits Exclusion or Unfair Regulatory Treatment	Passed Legislation that Promotes Affordable Housing, including Manufactured Housing	Court Decisions
California*; Colorado; Connecticut; Florida; Indiana; Iowa; Kansas*; Michigan; Minnesota*; Mississippi; Montana; Nebraska; New Hampshire; New Jersey; New Mexico; Oregon; Tennessee; Utah; Vermont*; Virginia	Hawaii; North Carolina; Illinois; Rhode Island; Washington	Georgia; Pennsylvania; Wisconsin

*State requires that manufactured and site-built housing be treated equally in regulations.

Source: Manufactured Housing Institutue

ENFORCEMENT ACTIVITIES

In recent years, HUD has taken steps to help ensure that manufactured housing is treated fairly and that local governments do not discriminate against this form of housing because it is built to a national code rather than state or local building codes. HUD has issued preemption letters and, in 1997, clarified its policy toward manufactured homes.

Preemption Letters

In 1992, HUD issued the first of several preemption letters to communities whose ordinances exclude manufactured homes on the basis that such homes were not built to local codes. The following is from a preemption letter issued by HUD's Office of Consumer and Regulatory Affairs to a local government in 1998:

> In 1975, pursuant to the National Manufactured Housing Construction and Safety Standards Act of 1974, the Department issued the Manufactured Home Construction and Safety Standards. Section 604(d) of the Act, 42 U.S.C. Section 5403(d), states:
>
>> Whenever a Federal manufactured home construction and safety standard established under [the Act] is in effect, no state or political subdivision of a state shall have any authority...to establish...with respect to any manufactured home covered, any standard regarding construction or safety applicable to the same aspect of performance of such manufactured home which is not identical to the Federal manufactured home construction and safety standard.
>
> In addition, the Manufactured Home Procedural and Enforcement Regulations [Regulations], 24 C.F.R. Section 3282.11(e), prohibit any state or locality from establishing and enforcing rules or taking any action that impairs Federal superintendence of the manufactured home industry.
>
> Generally, local ordinances regulating the location of manufactured homes have not been subject to the regulatory authority of the Act because such ordinances rest on the locality's right to determine proper land use. However, if a locality is attempting to regulate and even exclude certain structures that meet the Federal definition of manufactured home based solely on a construction and safety code different

In recent years, HUD has taken steps to help ensure that manufactured housing is treated fairly and that local governments do not discriminate against this form of housing because it is built to a national code rather than state or local building codes. HUD has issued preemption letters and, in 1997, clarified its policy toward manufactured homes.

than that prescribed by the National Manufactured Housing Construction and Safety Standards Act of 1974, the locality is without authority to do so.

Under Section 604(d) of the Act and the Regulations, a locality cannot accept manufactured homes meeting different standards and exclude or restrict manufactured homes that are aesthetically the same but only meet the Federal Standards.... To the extent that local zoning regulations require standards other than the Federal Standards for manufactured homes, they are preempted by the Act.

Clarification of HUD Policy

In the May 5, 1997, *Federal Register*, HUD clarified its policy on the application of the National Manufactured Housing Construction and Safety Standards Act and reemphasized the fact that local governments do not have the authority to exclude manufactured homes based solely on a construction and safety code different from that prescribed by the act. HUD further discusses the need to eliminate barriers "to the expanded use of affordable housing, including manufactured homes," explaining that this has been one of the primary objectives of the President's National Homeownership Strategy that attempts to identify and promote ways to increase homeownership and the supply of affordable housing. HUD coordinates the National Partners in Homeownership, which includes more than 60 national organizations, including the MHI and APA, and is charged with the responsibility of carrying-out the national strategy.

Part 2

Key Regulatory Issues

This section examines basic definitions and distinctions that should be made in zoning ordinance provisions for manufactured housing; appearance standards for manufactured homes; where and under what conditions manufactured housing may be permitted; and installation requirements.

DEFINITIONS AND DISTINCTIONS

Defining key terms and words in zoning ordinances serves to clarify and simplify the regulations. Local ordinance definitions of manufactured and other housing units built in a factory, however, continue to create confusion. Some state courts have overruled local regulations because of unclear definitions or because a critical distinction was not made between manufactured housing and other kinds of housing.

The previous PAS Reports on the subject have stressed the importance of distinguishing between the different types of homes built in a factory and the need for uniformity in the terms and definitions used. The current survey found that many communities continue to refer to manufactured homes, built since 1976 to the HUD Code, as mobile homes. A number of communities use the official designation, manufactured housing, as a general term referring to all types of housing built in a factory.

There are a number of different types of structures built in a factory. Some are indistinguishable from site-built housing in regards to appearance and durability, and clearly belong in residential neighborhoods with single-family detached site-built units. Included in this category of factory-built homes are a growing number of multisection and, more recently, two-story manufactured homes. Some factory-built units, on the other hand, may not be entirely appropriate in certain residential districts. How these different units are identified and subsequently defined will determine how they are regulated and, more importantly, determine whether they will be allowed in residential neighborhoods, restricted to special zoning districts, or prohibited altogether.

Factory-Built Housing

The zoning ordinance should include a term that provides a general classification for homes built in a factory. A useful general term is "factory built." Huntington, West Virginia, for example, defines the general term as follows:

> **factory-built housing**   A structure designed for long-term residential use. For the purpose of these regulations, factory-built housing consists of three types: modular, mobile homes, and manufactured homes.

This definition is nearly identical to one recommended in the 1986 PAS Report. It serves to both establish the general classification for homes built in a factory and also identifies the basic types of factory-built housing that communities will want to define in their ordinance:

1. Manufactured homes built in compliance with the HUD Code

2. Mobile homes, a designation that is retained to identify manufactured homes built before enactment of the HUD Code

3. Modular homes built to state or local requirements (panelized and other homes built to local codes could also be included in this category)

Manufactured homes. APA's 1985 survey found that only a few communities have followed the lead of the Federal Manufactured Home Construction and Safety Standards Act of 1974 and changed the term "mobile home" to "manufactured home" in their zoning ordinances. The 1996 survey found that many communities have updated their zoning provisions in recent years and now use "manufactured home." The 1996 survey also revealed that a growing number of communities refer to manufactured homes as "dwelling units" and use this designation in their definition of these homes, rather than referring to these homes only as "structures" or "factory-built structures" or anything else other than a dwelling. Where parity is practiced in the regulation of factory-built and site-built homes, manufactured homes are often included in the definition of a single-family home. Winnebago County, Wisconsin, for example, defines a single-family dwelling as follows:

> **single-family dwelling** A permanent structure placed on a permanent foundation, having one or more rooms, with provisions for living, sanitary, and sleeping facilities arranged for the use of one or more individuals of the same family. . . . These dwellings shall include site-built, manufactured, and modular homes.

APA's 1985 survey found that a number of the communities use the definition of a manufactured home included in the HUD Code as a guide in developing definitions in their zoning ordinances. The definition contained in the HUD Code, however, includes provisions that are more appropriately handled in a building code, such as minimum unit-size requirements. While using the HUD Code definition ensures a degree of consistency with the national code, the inclusion of minimum unit-size requirements in the definition is generally not considered to be good practice.

The current survey found that many communities have developed more concise and less wordy definitions that do not include dimensional requirements and recognize that the manufactured home is a dwelling unit. Huntington, West Virginia, has taken this approach to define the manufactured home and uses language that is nearly identical to that recommended by MHI and which appeared in the 1986 PAS Report.

> **manufactured home** A dwelling unit fabricated in an off-site manufacturing facility for installation or assembly at the building site, bearing a label certifying that it is built in compliance with the Federal Manufactured Housing Construction and Safety Standards Act of 1974 (42 USC 5401, et. seq.), which became effective June 15, 1976.

In 1997, MHI revised its recommended definitions for factory-built housing. The new definitions are short and simple, and each emphasizes the fact that these homes are residential uses. MHI's new definition for manufactured housing illustrates its current approach.

> **manufactured home** A residential dwelling built in accordance with the Federal Manufactured Home Construction and Safety Standards.

The 1996 survey also revealed that a growing number of communities refer to manufactured homes as "dwelling units" and use this designation in their definition of these homes, rather than referring to these homes only as "structures" or "factory-built structures" or anything else other than a dwelling. Where parity is practiced in the regulation of factory-built and site-built homes, manufactured homes are usually included in the definition of a single-family home.

Mobile homes. MHI recommends the following streamlined definition for a mobile home—a term that should be retained in the zoning ordinance to identify pre-HUD Code homes.

> **mobile home** A residential dwelling that was fabricated in an off-site manufacturing facility, designed to be a permanent residence, built prior to enactment of the Federal Manufactured Home Construction and Safety Standards.

Modular homes. Modular housing is defined in local ordinances so that it and other factory-built homes that conform to local or state building codes are distinguished from manufactured homes that are built to a national building code. The definition of modular housing from the Montgomery County, Maryland, ordinance describes this type of housing by the code that governs it and is typical of how many of the communities responding to the current survey define this type of factory-built housing.

> **modular home** A structure intended for residential use and manufactured off-site in accord with the [local or state] BOCA Basic Building Code.

In the past, modular homes had been further distinguishable from manufactured homes by their appearance (they more closely resembled site-built homes) and overall durability. Improvements in the appearance and quality of manufactured homes, however, and the more recent practice of stacking manufactured housing modules to create two-story homes have made these distinctions less prominent. A number of manufactured housing firms now build modular homes and, in some cases, build them in manufactured housing plants.

Manufactured home development. The 1986 PAS Report recommended that definitions for different types of manufactured housing developments may be "derived like those for factory-built housing—begin with a general classification and then refine by specific type of development." So, an appropriate definition would look something like this:

> **manufactured home development** A general category of development that includes manufactured home subdivisions and manufactured home communities (or land-lease communities).

This definition includes the two basic types of manufactured home developments that are further defined below. The term "communities" replaces the term "parks" that appeared in the earlier report and is still used in many ordinances. A growing number of local governments, however, no long refer to these residential developments as parks, and, similarly, the manufactured housing industry has for some time recommended that manufactured home parks be referred to as "land-lease developments."

Referring to these developments as communities recognizes that they are residential land uses that belong in residential districts. Unlike the earlier mobile home parks, which often lacked many of the amenities included in more conventional residential developments, many modern manufactured home communities include improved common open space for outdoor activities, a community building, and playgrounds. Some may even include on-site day care facilities and after-school programs.

> **manufactured home subdivision** A subdivision designed and/or intended for the sale of lots for siting manufactured homes.

In many cases, a manufactured housing subdivision is subject to the same land development and site-improvement standards that apply to subdivisions containing site-built homes. Individual lots in a manufactured housing subdivision are sold in the same manner as lots for site-built homes in conventional subdivisions.

The term "communities" replaces the term "parks" that appeared in the earlier report and is still used in many ordinances. A growing number of local governments, however, no long refer to these residential developments as parks, and, similarly, the manufactured housing industry has for some time recommended that manufactured home parks be referred to as "land-lease developments."

In many cases, a manufactured housing subdivision is subject to the same land development and site-improvement standards that apply to subdivisions containing site-built homes. Individual lots in a manufactured housing subdivision are sold in the same manner as lots for site-built homes in conventional subdivisions.

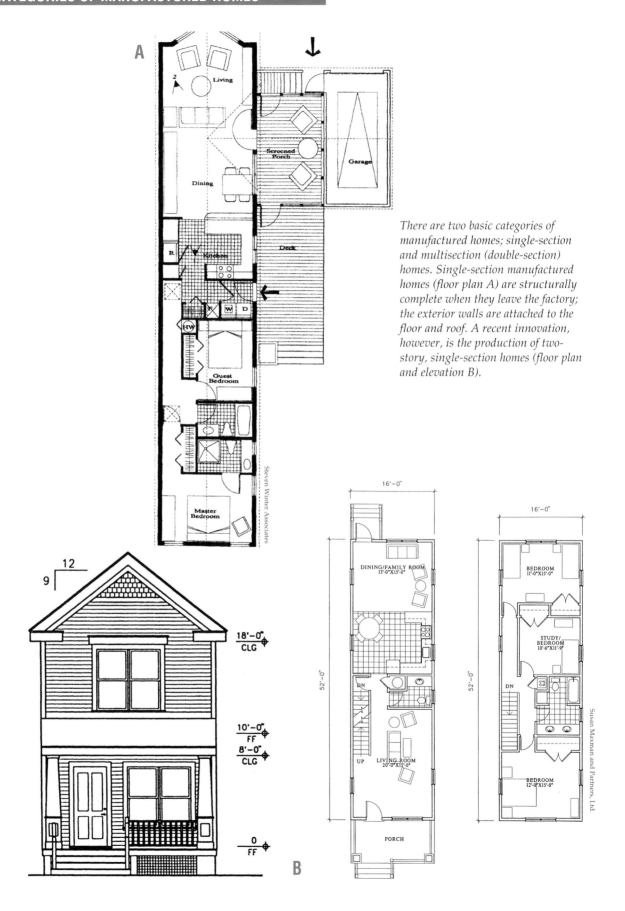

There are two basic categories of manufactured homes; single-section and multisection (double-section) homes. Single-section manufactured homes (floor plan A) are structurally complete when they leave the factory; the exterior walls are attached to the floor and roof. A recent innovation, however, is the production of two-story, single-section homes (floor plan and elevation B).

14

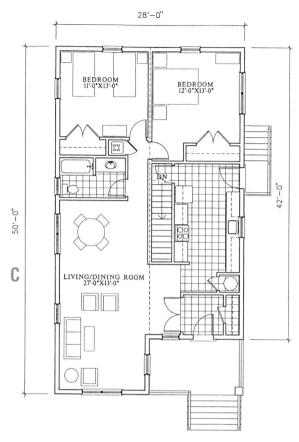

28'-0"

50'-0"

42'-0"

BEDROOM
11'-0"X13'-0"

BEDROOM
12'-0"X13'-0"

8'-0"
CLG

0
FF

DN

LIVING/DINING ROOM
27'-0"X13'-0"

C

12
11

Multisection homes (floor plan and elevation C) consist of two or more components, each with three walls attached to a roof and floor section. These components are joined at the side. Multisection components can now be stacked to create two-story homes (floor plan and elevation D). Each of these types of manufactured homes can can be enhanced with the addition of site-built porches and other elements.

Susan Maxman and Partners, Ltd.

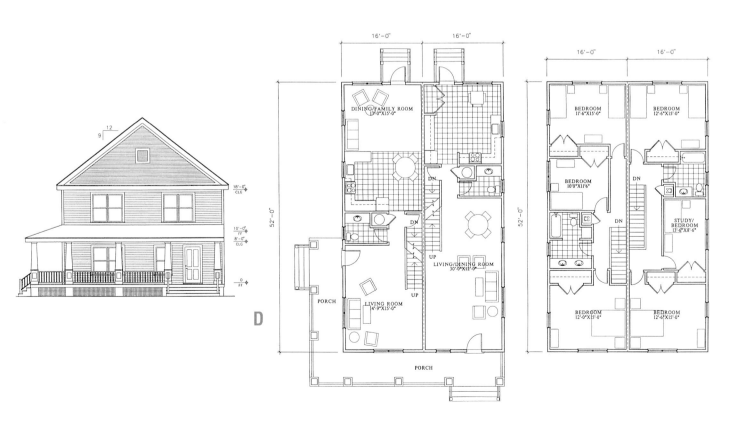

16'-0"   16'-0"

52'-0"

DINING/FAMILY ROOM
13'-0"X15'-0"

DN

DN

UP

LIVING/DINING ROOM
30'-0"X12'-0"

UP

PORCH

LIVING ROOM
14'-9"X15'-0"

D

18'-0"
CLG

10'-0"
FF
8'-0"
CLG

0
FF

12
9

PORCH

16'-0"   16'-0"

52'-0"

BEDROOM
11'-6"X15'-0"

BEDROOM
12'-6"X15'-0"

BEDROOM
10'0"X11'6"

DN

STUDY/
BEDROOM
13'-0"X8'-6"

DN

BEDROOM
12'-0"X15'-0"

BEDROOM
12'-6"X15'-0"

15

The City of Issaquah, Washington, includes the following definition for a manufactured home community (land-lease development) in its zoning ordinance:

> **manufactured housing community** Any piece of real property under single ownership or control for which the primary purpose is the placement of two or more manufactured homes for permanent residential dwellings and for the production of income. A manufactured housing community does not include real property used for the display and sale of manufactured units, nor does it include real property used for seasonal recreational purposes only, as opposed to year-round occupancy.

This definition addresses a number of concerns that local officials may have in regards to manufactured housing communities, including the display and sale of manufactured homes and seasonal recreational uses. These concerns may be more appropriately addressed in a section of the zoning ordinance that establishes specific requirements for manufactured housing.

The 1986 PAS Report recommended the following definition for manufactured home communities or land-lease developments:

> **manufactured housing community** A parcel of land under single ownership on which two or more manufactured homes are sited.

MHI's updated definition is more detailed:

> **land-lease community** A residential development typified by single ownership of the land within the development, with the landowner retaining the rights of ownership. Home sites within the community are leased to individual homeowners, who retain customary leasehold rights.

This definition establishes the major distinction between a manufactured home community or land-lease community and a manufactured home subdivision—individual lots or spaces in a manufactured home community are rented and not owned. Dwelling units in a land-lease community are usually taxed as personal property, whereas units in subdivisions are taxed as real property, like site-built homes.

Most communities require that manufactured home communities include a certain amount of common open space and/or common facilities for residents. Subdivision developments, on the other hand, are not usually required to set aside a portion of the development for common open space, but a certain amount of land may have to be dedicated for public use or a fee must be paid in lieu of such dedication.

Some communities permit cooperative and condominium ownership of manufactured home developments as an alternative to the land-lease arrangement. A cooperative title means that the land is held by a corporation in which each homeowner holds a share. Building sites are leased to individual shareholders, who may sell or transfer their leases. The title of each individual site, however, remains with the corporation. In a condominium development, residents own their individual site, and an association of homeowners owns the common areas of the development.

If there is a need to distinguish between manufactured home condominium or cooperative developments and other types of manufactured home development, the following definition might be used.

> **manufactured home condominium (or cooperative)** A condominium (or cooperative) development of manufactured homes.

Most communities require that manufactured home communities include a certain amount of common open space and/or common facilities for residents. Subdivision developments, on the other hand, are not usually required to set aside a portion of the development for common open space, but a certain amount of land may have to be dedicated for public use or a fee must be paid in lieu of such dedication.

APPEARANCE STANDARDS

The appearance of manufactured housing continues to improve, and some manufactured homes are indistinguishable from site-built homes. The industry still produces single-section units with long narrow bodies and a 3:12 roof pitch, but most manufactured homes produced today are larger multisection units that more closely resemble site-built homes. In addition, the appearance of some single-section units has been enhanced with attached garages and porches, and, for the first time, single-section modules are being stacked to create two-story houses to give manufactured housing a whole new look.

Most communities have not been too concerned about the appearance of manufactured housing when it is placed in a manufactured home development or located in remote rural areas. When these homes are permitted outside such developments and in residential districts, communities use appearance standards to help ensure compatibility with site-built homes. Most have established appearance standards for manufactured homes only. Some have established appearance standards for all types of single-family detached housing whether it is built in a factory or on site. As noted in Part One, nearly a third of the communities responding to the current survey indicated that their zoning ordinance did not include specific regulatory provisions for manufactured housing that did not also apply to site-built housing.

Standards For Manufactured Homes Only

Over the past decade or so, little has changed in the way most communities regulate the appearance of manufactured homes. The appearance standards that apply only to manufactured housing still vary widely. Some communities require only that the exterior finish of the unit, typically siding and roofing, must consist of materials customarily used in site-built housing. A number of communities, however, have established more elaborate appearance standards that include architectural guidelines with specific standards for roof pitch and overhang; type of exterior materials that can be used; and minimum allowable size when a manufactured house is sited on an individual lot in a residential district. Sandusky, Ohio, for example, has established the following design criteria for manufactured homes that are permitted in residential districts:

1. **Exterior Siding.** Exterior siding shall be made of nonreflective and nonmetallic materials unless approved by the City Engineer. Acceptable siding materials include: vinyl, wood, stucco, brick, stone, or other masonry materials or any combination of these materials, or any material that the City Engineer deems to meet the intent of this code.

2. **Color/Texture.** Color and texture of exterior materials shall be compatible with the adjacent single-family structures.

3. **Roof Structure.** Except for authorized deck areas, all roof structures should be sloped and provide an eave projection of no less than six inches and no greater than 30 inches.

4. **Roofing Material.** All roofing material shall consist of one of the following categories: wood, shingle, wood shake, synthetic composite shingle, concrete tile, or any other material that the City Engineer deems to meet the intent of this code. Metallic roofing surfaces shall not be permitted on the residential structure or on any garage or carport unless approved by the City Engineer.

5. **Minimum Floor Area.** The minimum floor area for every dwelling located on a lot in a residential zoning district, which is not part of a [manufactured housing] subdivision, shall be 800 square feet, excluding the garage or carport.

When these homes are permitted outside manufactured housing developments and in residential districts, communities use appearance standards to help ensure compatibility with site-built homes. Most have established appearance standards for manufactured homes only. Some have established appearance standards for all types of single-family detached housing whether it is built in a factory or on site.

The appearance standards that apply only to manufactured housing still vary widely. Some communities require only that the exterior finish of the unit, typically siding and roofing, must consist of materials customarily used in site-built housing. A number of communities, however, have established more elaborate appearance standards that include architectural guidelines with specific standards for roof pitch and overhang; type of exterior materials that can be used; and minimum allowable size when a manufactured house is sited on an individual lot in a residential district.

17

6. **Minimum Width.** The minimum width of a dwelling unit located on a lot outside of a [manufactured housing] subdivision shall be 20 feet.

7. **Foundation**. All manufactured homes shall be placed on a permanent foundation that meets applicable building code requirements of the City of Sandusky, such that the floor elevation of the proposed dwelling is reasonably compatible with the floor elevations of surrounding dwelling units.

These provisions were among the most extensive identified. Note that some flexibility in their application is allowed. For example, specific acceptable siding materials are identified in the provisions, but "any material that the City Engineer deems to meet the intent of this code" may also be used. These provisions are also less restrictive than many others in their treatment of the design of the roof. Most communities that regulate this aspect of manufactured housing establish a minimum roof pitch requirement. Sundusky's provisions only requires that roof structures are "sloped."

Huntington, West Virginia, has also established a substantial list of appearance standards for manufactured housing–standards that also apply to modular housing. The city requires that manufactured and modular homes that are located in residential districts have a width and length of at least 22 feet, specific siding, floor area, minimum roof pitch of 2:12 (two feet in rise for every 12 feet of horizontal run). The city does, however, allow for deviations from these requirements:

> The building official may approve deviations from one or more of the developmental or architectural standards provided herein on the basis of finding that the materials to be utilized or the architectural style proposed for the dwelling will be compatible and harmonious with existing structures in the vicinity.

A simpler and seemingly less restrictive approach has been taken by Riley County, Kansas, which has established specific standards for "Residential-Design Manufactured Homes," requiring a minimum width of 22 feet, a roof pitch of 2.5:12, solid concrete or reinforced concrete block perimeter foundation, and "siding and roofing materials which are customarily used on site-built homes." The minimum unit-width provision in this otherwise reasonable set of standards prohibits the use of single-section homes on individual lots in residential districts, regardless of how the home is designed or whether it includes a porch or attached garage that can allow for a conventional appearance. The minimum width requirement would also prohibit the use of stacked single-section homes.

What may be the most reasonable approach (and one that a growing number of communities are embracing) is to replace the specific appearance standards with a general requirement that the appearance of manufactured units located in residential districts must be similar to that of other dwelling units in the district. For example, Moreno Valley, California, requires that the character, appearance, and certain materials used in manufactured units must be similar to what is used in other dwelling units in the area.

1. The design of the structure shall be similar in character and appearance to other dwellings in the area with regard to unit size, roof overhangs, roof materials, roof pitch, and exterior materials.

2. All building setbacks, parking, coverage, height, width and sign requirements of the base district shall apply.

3. A roof constructed of asphalt composition, shingle, tile, crushed rock, or similar roofing material (except metal), which is compatible with surrounding development, [shall be used].

What may be the most reasonable approach (and one that a growing number of communities are embracing) is to replace the specific appearance standards with a general requirement that the appearance of manufactured units located in residential districts must be similar to that of other dwelling units in the district.

4. Exterior siding of brick, wood, stucco, plaster, concrete, or other material, which is finished in a nonglossy and nonreflective manner and which is compatible with surrounding development [shall be used].

5. A predominant shape and form that is compatible with the surrounding neighborhood [shall be used].

6. If an enclosed garage is required within the zoning district in which the dwelling unit is to be located, the design and materials of the garage shall be compatible with the main dwelling.

Standards for Both Manufactured and Site-Built Homes

Appearance standards that are applied equally to both site-built and manufactured homes usually include provisions regulating exterior finish and roof construction, but are more likely to also include a minimum unit-width requirement—ranging from 20 to 24 feet. (The 1985 survey found that a range of 18 to 22 feet was more typical of provisions for both manufactured and site-built homes.) In some cases, only a minimum width requirement is imposed to ensure that the appearance of the manufactured home is compatible with that of the typical site-built home. Polk County, Iowa, for example, requires only that:

> The minimum dimension of the main body of the [single-family detached] dwelling unit shall not be less than 22 feet.

Polk County does, however, require that a factory-built unit be taxed as real property if it is classified as a "single-family dwelling unit" in the same sense as a site-built unit.

In most cases, the minimum unit-width requirement is paired with either a minimum unit-size or minimum roof pitch requirement. Winnebago County, Wisconsin, requires:

> One-family dwellings shall contain a minimum width of 20 feet, measured from the narrowest part of the structure, and a minimum area of 1,000 square feet.

Cottage Grove, Minnesota, has established both minimum width and roof pitch requirements for single-family dwellings, which also means manufactured homes, and includes unit placement and foundation requirements in its provisions.

1. **Minimum width.** The minimum width of the main portion of the structure shall be 20 feet, as measured across the narrowest portion.

2. **Minimum roof pitch.** The pitch of the main roof shall be not less than 2.5 feet of rise for each 12 feet of horizontal run. The requirement may be waived for earth-sheltered structures.

3. **Placement.** Every single-family dwelling shall be placed so that the apparent entrance or front of the home faces or parallels the principal street frontage, except where the lot size exceeds one acre.

4. **Foundations.** Every dwelling shall be placed on a permanent foundation in compliance with the Uniform Building Code as adopted by the city.

Almont Township, Michigan, has established minimum width, roof pitch, and storage area requirements for single-family dwellings but allows for a variance from the minimum width and roof pitch requirements.

1. All such units shall have a minimum width on the narrowest side of 23 feet. Where the architectural style proposed includes something less than 23 feet, a variance must be obtained from the Zoning Board of Appeals.

Appearance standards that are applied equally to both site-built and manufactured homes usually include provisions regulating exterior finish and roof construction, but are more likely to also include a minimum unit-width requirement—ranging from 20 to 24 feet.

2. All housing units shall have a roof with a minimum of 3:12 pitch and 12-inch overhang on all sides of the roof. Where the architectural style of the unit incorporates a flat roof, a pitch of less than 3:12, or no overhang, a variance must be obtained from the Zoning Board of Appeals.

3. All single-family dwellings shall have a minimum storage area of 100 square feet, exclusive of the required minimum square footage of the zoning district. The storage area may be located as part of the dwelling, in a basement area, as part of a garage, or in a separate storage building constructed at the same time as the dwelling unit.

4. All single-family dwellings shall meet the minimum lot area, minimum setbacks, maximum height limitations, and minimum floor area requirements for the particular district in question. Off-street parking shall be provided according to Section 6.04.

5. Any single-family dwelling that was not specifically designed for placement on an approved foundation shall have cable tie-downs installed at least every six feet prior to occupancy to protect the unit from windstorm damage. All units shall be provided with a perimeter wall of the same perimeter dimensions of the dwelling unit, whether bearing or otherwise, and constructed of such material and type as required in the applicable Building Code for single-family dwellings. Wood basements may be permitted, provided that they comply with the requirements of the Township Building Code.

6. The building shall have all towing apparatus, wheels, and exposed chassis removed before occupancy of any kind is permitted.

Provo City, Utah, has established one of the more elaborate collections of appearance and design standards, and is the only community—among those singled out for discussion in this section—that has included provisions regulating exterior finish of single-family dwellings. The city does, however, allow for deviations from any one or more of these provisions.

1. The dwelling must meet the requirements of the Provo City Building Code, or, if it is a manufactured home, it must meet the requirements of the HUD Code and must not have been altered in violation of such codes. A used dwelling must be inspected by the Chief Building Official or his designee prior to placement on a lot to insure it has not been altered in violation of such codes. Any violations must be corrected as directed by the Chief Building Official.

2. The dwelling must be taxed as real property. If the dwelling is a manufactured home, an affidavit must be filed with the State Tax Commission pursuant to Utah Code Annotated 59-2-602.

3. The dwelling must be approved for and permanently connected to all required utilities.

4. Each dwelling shall have a code-approvable, site-built, concrete, masonry, steel, or treated wood foundation capable of transferring design dead loads and live loads and other design loads unique to local home sites due to wind, seismic, and water conditions that are imposed by or upon the structure into the underlying soil or bedrock without failure. All foundations shall be designed in accordance with Provo City adopted building codes, the manufacturer's recommendations, NCSBCS Standards, or an approved engineered design. All perimeter footings must be a minimum of 30 inches below grade for frost protection. All tie-down devices must meet Provo City adopted building codes, the manufacturer's recommendations, NCSBCS Standards, or an approved engineered design. The space beneath the structure must be enclosed at the perimeter of the dwelling in accordance with the manufacturer's recommendations

Provo City, Utah, has established one of the more elaborate collections of appearance and design standards, and is the only community—among those singled out for discussion in this section—that has included provisions regulating exterior finish of single-family dwellings.

or NCSBCS Standards and constructed of materials that are weather resistant and aesthetically consistent with concrete or masonry type foundation materials. All manufactured home running gear, tongues, axles, and wheels must be removed at the time of installation.

5. Dwelling shall have a roof surface of wood shakes, asphalt, composition, wood shingles, concrete, fiberglass or slate tiles, or built-up gravel materials. Unfinished galvanized steel or unfinished aluminum roofing shall not be permitted. There shall be a roof overhang at the eaves and gable ends of not less than six inches, excluding rain gutters, measured from the vertical side of the dwelling. The roof overhang requirement shall not apply to areas above porches, alcoves, and other appendages which together do not exceed 25 percent of the length of the dwelling.

6. Dwellings shall have exterior siding materials consisting of wood, hardwood, brick, concrete, stucco, glass, metal or vinyl lap, tile, or stone.

7. The width of the dwelling shall be at least 20 feet at the narrowest point of its first story for a length of at least 20 feet exclusive of any garage area. The width shall be considered the lesser of the two primary dimensions. Manufactured homes shall be multiple transportable sections.

8. Single dwellings in the R-1-10 zone shall be provided with a garage or carport [with] a minimum interior width of 20 feet and constructed concurrently with the dwelling. Single dwellings in all other zones shall be provided with a garage or carport having a minimum interior width of 12 feet.

9. **Minimum Floor Area.** All single-family detached dwellings shall have a minimum floor area on the main floor (exclusive or garage) as follows:

> R-1-6: 850 SF
>
> R-1-7: 950 SF
>
> R -1-8: 1,000 SF
>
> R -1-9, 10, 15, 20: 1,200 SF
>
> R-2: 900 SF single-family; 750 SF per unit in duplex
>
> R-2.5, 3, 4, 5 and A-1: 900 SF
>
> R-A: 1,000 SF

10. Wood or metal porches, decks, or verandas are only permitted on the front of the home when covered with a roof.

11. The Community Development Director may approve deviations from one or more of the developmental or architectural standards contained herein in subsections 5 through 10 on the basis of a finding that the architectural style proposed provides compensating features and that the proposed dwelling will be compatible and harmonious with existing structures in the vicinity. The determination of the Community Development Director may be appealed to the Board of Adjustment pursuant to the provisions of Section 14.02.050.

Fairness and Flexibility

Communities are using a variety of standards to enhance the appearance of manufactured housing and to help ensure that these homes are compatible with site-built homes. The scope of these standards has, in many cases, been dictated by state law, with some states allowing only limited application of appearance standards to manufactured homes and a few states (see Table 3) calling for parity in the regulation of manufactured homes and site-built homes.

Communities are using a variety of standards to enhance the appearance of manufactured housing and to help ensure that these homes are compatible with site-built homes. The scope of these standards has, in many cases, been dictated by state law, with some states allowing only limited application of appearance standards to manufactured homes and a few states calling for parity in the regulation of manufactured homes and site-built homes.

Closer examination of the parity approach reveals that it may not be as reasonable as it may initially appear to be. As mentioned previously, communities that have established appearance standards that apply to both manufactured homes and site-built housing are more likely to include a minimum width requirement that is typically about 20 feet. This requirement excludes all single-section manufactured homes regardless of how they may otherwise look, how they are sited, whether they have been enhanced with a garage or other additions, or are stacked.

Whether appearance standards and other design requirements apply only to manufactured homes or all types of single-family homes, it is important to consider the subsequent impact that these standards will have on development. Some standards can discourage innovation and imagination in unit and site design, others can add unnecessary costs, and some can effectively exclude manufactured housing altogether.

Closer examination of the parity approach reveals that it may not be as reasonable as it may initially appear to be. As mentioned previously, communities that have established appearance standards that apply to both manufactured homes and site-built housing are more likely to include a minimum width requirement that is typically about 20 feet. This requirement excludes all single-section manufactured homes regardless of how they may otherwise look, how they are sited, whether they have been enhanced with a garage or other additions, or are stacked.

Communities that believe a minimum width and other quantitative or specific requirements should be in place may consider allowing exceptions to be made when appropriate. A number of the communities allow for some deviation from the established appearance standards for manufactured housing when it is determined that "the architectural style proposed provides compensating features and. . .the proposed dwelling will be compatible and harmonious with existing structures in the vicinity" (Provo City, Utah). Such an exception could be made, for example, to allow single-section homes in a neighborhood of smaller homes and/or narrow lots. Stacked single-section modules and single-section units with attached garages may also be appropriate in certain residential districts. Two of the manufactured home developments featured in Part 4 of this report use single-section manufactured homes that have been designed for infill development.

A growing number of communities do not use specific requirements like minimum width standards. Instead, standards mandate that the appearance of manufactured homes sited in residential districts should be similar and/or compatible with site-built homes in the area. Nearly a third of the communities responding to the 1996 survey of PAS subscribers indicated that they have not established any special appearance requirements for manufactured housing. In most cases, manufactured dwelling units are subject only to the same site development standards—such as building setback, height, lot coverage, and so on—that apply to site-built dwelling units. A growing number of communities are taking this more flexible approach to regulating manufactured housing.

Whether appearance standards and other design requirements apply only to manufactured homes or all types of single-family homes, it is important to consider the subsequent impact that these standards will have on development. Some standards can discourage innovation and imagination in unit and site design, others can add unnecessary costs, and some can effectively exclude manufactured housing altogether. For example, requiring that the longest dimension of a manufactured unit be sited parallel to the street can prohibit the use of innovative and more cost-effective site designs like zero lot line and clustering. Typically, zero lot line plans require that the longest dimension of the home be sited on the side lot line in order to take full advantage of small lots. Similarly, clustering units is difficult, if not impossible, when every unit has to be sited parallel to the street.

One of the communities studied, Issaquah, Washington, permits zero lot line siting of dwelling units and encourages cluster subdivision to allow for more affordable housing. Manufactured housing subdivisions are subject to the same land development and site improvement standards that apply to conventional subdivisions, but the city encourages that manufactured home subdivisions be designed as cluster developments.

> **Cluster encouraged.** A cluster development within a manufactured home subdivision shall be encouraged in order to provide affordable housing through the provision of smaller lots. A cluster development is encouraged to have 15 percent usable open space, which does not include critical areas or their required buffers. This usable open space should have the ability to provide for recreation.

Appearance standards should be flexible, their impact on subsequent development should be considered, and these standards should be tailored to local circumstances. Appearance standards that apply to all types of single-family homes are generally fairer than those that apply to only one type of housing. Minimum unit-width and certain siting requirements can exclude some types of manufactured housing and make it difficult to take advantage of innovative and cost-effective site designs. An encouraging trend is increasing use of appearance provisions requiring that manufactured housing is similar to and compatible with existing housing, rather than requiring that manufactured housing adheres to a range of specific, quantitative standards.

Appearance standards have helped to improve the appearance of manufactured housing. The extent to which these standards add to the cost of manufactured homes should be considered since affordability is still the most attractive feature of the manufactured home.

Location and Development Options

More communities than ever before are allowing manufactured housing by right in residential districts and outside of manufactured home developments on individual lots when certain unit design and siting requirements are met. As noted previously, since 1986, the increase in the number of communities reporting that they permit manufactured units on individual lots in existing residential districts has been dramatic. Fewer than a fifth of the communities that allow manufactured housing in residential districts do so under special review procedures; over four-fifths allow them by right. Ten years ago, nearly half of the communities that allowed manufactured homes in residential districts required that development proposals undergo special review.

Following the lead of the majority of the communities that responded to the current survey suggests that the ideal permitting system for manufactured housing would:

- permit manufactured housing by right in existing residential districts and outside of manufactured housing developments when certain unit design and siting requirements are met;

- restrict manufactured housing to less-restrictive residential districts or manufactured home developments when these established requirements are not met and less stringent standards apply;

- require that the appearance of manufactured housing is similar and compatible with that of site-built housing when it is allowed in residential districts;

- allow some flexibility and, when appropriate, some exceptions to those appearance standards when the appearance of the proposed manufactured housing is compatible with the appearance of site-built homes in the area; and

- establish a single set of unit design and siting standards for site-built and manufactured housing.

What development options make sense for manufactured housing? The 1986 PAS Report provided the following commentary on the subject that is worth restating here:

> Ideally, developers should be able to choose the regulatory approach that best suits their development proposal. Of course, there is nothing innovative about this approach to land development—site-built hous-

Appearance standards have helped to improve the appearance of manufactured housing. The extent to which these standards add to the cost of manufactured homes should be considered since affordability is still the most attractive feature of the manufactured home.

Fewer than a fifth of the communities that allow manufactured housing in residential districts do so under special review procedures; over four-fifths allow them by right.

ing has been developed in this manner for some time. A range of development options allows the scale and location of a proposed development, as well as other factors, to be taken into account in determining the best regulatory technique to use. Large-scale developments with small-lots, high densities, innovations in unit siting, and common open space and facilities are best regulated under planned unit development provisions. On the other hand, more conventional, large-lot manufactured home developments that do not provide common open space can be adequately handled by standard subdivision procedures.

In addition, as noted above, the cluster development option could also be offered as an alternative to conventional subdivision design.

Unit Installation Requirements

The HUD Code does not include site-specific provisions for the installation of a manufactured home on a site, nor does it specify the manner in which the individual sections of multisection homes are to be joined together on the site. The home manufacturer is required to provide installation instructions, and the inspection and approval of installation of the home on a foundation are left to state and local building code administration.

Local officials should require that all manufactured units be installed on a "properly engineered" foundation system that meets the manufacturer's installation requirements and applicable state and local regulations. MHI defines a properly engineered foundation as:

> A foundation system that provides adequate support of the home's vertical and horizontal loads and transfers these and other imposed forces, without failure, from the home to the undisturbed ground (below the frost line in frost-susceptible areas).

About half of the states have adopted installation standards and a number of communities have adopted installation requirements established by the Federal Emergency Management Agency so that homeowners are eligible for flood insurance. Local officials and developers who need more information on installation requirements can consult the HUD publication, *Permanent Foundations Guide for Manufactured Housing* (see Appendix A for full citation), and/or contact the Manufactured Housing Association in their state (see Appendix E).

About half of the states have adopted installation standards and a number of communities have adopted installation requirements established by the Federal Emergency Management Agency so that homeowners are eligible for flood insurance. . . . Local officials and developers that need more information on applicable installation requirements should contact the Manufactured Housing Association in their state.

Part 3

Development Standards

Manufactured housing developments should be permitted in residential districts where residential uses of similar density and intensity are permitted. Manufactured housing subdivisions should be permitted in single-family districts where site-built subdivisions of comparable densities are permitted. Manufactured housing land-lease communities should be allowed in districts where site-built rental developments are permitted and can also be permitted in more-restrictive zoning districts if steps are taken to ensure compatibility with existing development or other types of development permitted in the district.

The development standards for manufactured housing developments, including both land-lease and subdvision developments, of 20 communities are examined in this section of the report. These communities were chosen from the respondents to the 1996 survey. They represent as complete a geographic cross-section and demographic mix as possible. Communities drafting new provisions or revising their existing zoning and subdivision requirements should find some helpful guidance in the information presented in Tables 4 through 11.

First, we examine the zoning districts where manufactured home developments are permitted and the conditions under which these developments are permitted. Specific land development standards are examined next, including those for: density, minimum lot size, and other lot dimensions; unit setback and separation; open space, perimeter, and landscaping; streets and sidewalks; and parking. The standards examined here are compared to those noted in the earlier 1986 report to determine if and how local standards for manufactured housing developments have changed over the past decade.

Slightly more than half of the communities featured in this section have established development standards that apply only to manufactured housing land-lease or rental communities. When this is the case, manufactured housing subdivisions are usually developed under the subdivision requirements that apply to site-built housing. Other communities featured here have established two different sets of development standards—a set for land-lease communities and a set for manufactured housing subdivisions.

LOCATION AND PERMITTING REQUIREMENTS

More than half of the 20 communities examined in this section have created a special zoning districts for manufactured housing developments. (See Table 4.) Springfield, Missouri, however, is the only community featured here that restricts all manufactured units to a manufactured home district. Most communities permit manufactured housing in existing residential districts if certain standards are met. (See the discussion in Part 2 about design standards for manufactured homes.) In most cases, manufactured housing districts are primarily reserved for land-lease communities and/or developments that include units that do not meet established standards that would otherwise allow these homes to be located in the same residential districts as site-built housing. Manufactured housing developments are usually permitted by right in manufactured housing districts, but a rezoning is often required to create the district.

Manufactured housing developments should be permitted in residential districts where residential uses of similar density and intensity are permit-

Table 4. Permitting Procedures

Jurisdiction	Where and How Manufactured Homes Are Permitted
Manufactured Home Developments:	
Albany, Oregon	Manufactured housing land-lease developments are permitted under site plan review approval in most residential districts.
	Manufactured housing developments may also be developed as a Planned Development where both homes and sites are individually owned and other lands and facilities are maintained and owned in common.
	Manufactured units are permitted on individual lots outside of manufactured housing developments when certain compatibility standards are met.
Hampton, Virginia	Manufactured housing developments are permitted in two multifamily districts by use permit. Manufactured units are restricted to manufactured housing developments.
Inver Grove Heights, Minnesota	Manufactured housing land-lease developments are permitted by conditional permit. Manufactured units are permitted in all residential districts when they meet the definition for a "one-family detached dwelling."
Poplar Bluff, Missouri	Manufactured housing subdivisions are permitted by right in two residential districts, RS-4 and RS-5. Land-lease developments are permitted as a conditional use in RS-5 residential districts.
Portland, Maine	Manufactured housing developments are permitted in flexible housing zone, subject to review by the planning board.
Roanoke County, Virginia	Manufactured housing developments are permitted in some residential districts; multisection homes are permitted in all residential districts. A manufactured home community may be developed as a land-lease community or as a condominium.
	Manufactured units are permitted by right in R-1 and other residential districts on individual lots.
Scappoose, Oregon	Manufactured housing developments require site development review. Manufactured homes are permitted in all residential districts.
West Sacramento, California	Manufactured housing developments are allowed in residential districts by use permit. Manufactured units are permitted by right in residential districts.

Table 4. Permitting Procedures (continued)

Jurisdiction	Where and How Manufactured Homes Are Permitted
Manufactured Home Development Districts:	
Brookings, South Dakota	Manufactured homes are permitted in one high-density district and the Residence RMH Single-Family and Manufactured Housing District.
Burnsville, Minnesota	Manufactured housing is permitted in all residential districts if certain requirements are met; otherwise it is restricted to the R-3D-Manufactured Housing District.
Charlotte-Mecklenburg County, North Carolina	Manufactured housing that meets certain design standards is permitted in all residential districts as an "overlay." Units that do not meet these standards are allowed only in the R-MH-Manufactured Housing District.
Grand Island Hall County, Nebraska	Manufactured housing is permitted in all residential districts if it meets standards established for "dwelling units"; otherwise it is permitted in the M and MD Manufactured Home (Overlay) Zones.
Hays, Kansas	Manufactured housing is permitted in all residential districts if it is a "Residential Design Manufactured Home"; otherwise it is restricted to M-P Manufactured Home Park (land lease), Residential District, or M-S Manufactured Home Subdivision Residential Districts.
Johnson County, Kansas	Manufactured housing in permitted in all residential districts if it is built as a "Residential-Design Manufactured Home"; otherwise units are restricted to Planned Residential Manufactured Home Park District (PRMHP) (land lease) or the Planned Residential Manufactured Home Subdivision District (PRMHS).
Lancaster, Texas	Manufactured homes are permitted in residential districts by special use permit and by right, and in developments in the MH-1, Modular Home District.
Owensboro-Daviess County, Kentucky	Class A manufactured homes are permitted in residential districts and others in Planned Manufactured Housing Park-MHP Zone (land lease).
Pinellas County, Florida	"Residential Design Manufactured Homes" are permitted in all residential districts. Units not meeting the standards established for residential-designed homes and developments are restricted to the R-6 Residential Districts for land-lease and subdivision developments.
Springfield, Missouri	Manufactured housing is permitted in only the R-MHC-Manufactured Home Community District.
Wichita-Sedgwick County, Kansas	Manufactured housing is permitted in residential districts if it is a "Residential Design Manufactured Home"; otherwise, such housing is restricted to MH-Manufactured Housing District for both land-lease and subdivision developments.
Winston Salem-Forsyth County, North Carolina	Manufactured housing is permitted in the Manufactured Home Development (MH) District under review by the planning board.
	Manufactured homes are permitted in residential districts. The county has established "improvement" standards for existing manufactured housing developments.

ted. Manufactured housing subdivisions should be permitted in single-family districts where site-built subdivisions of comparable densities are permitted. Manufactured housing land-lease communities should be allowed in districts where site-built rental developments are permitted and can also be permitted in more-restrictive zoning districts if steps are taken to ensure compatibility with existing development or other types of development permitted in the district. This all made sense a decade ago when the 1986 report on the subject was prepared, and there may even be more support for this thinking today. Continued improvement in the appearance and quality of manufactured housing alone is reason enough to treat manufactured housing as a residential use that belongs in residential districts.

If a community is going to establish a special district for a manufactured housing development, an ample supply of land suitable for residential development should be zoned for such a district in advance of development proposals. Otherwise, such an area may require rezoning, which can be expensive, adding significantly to the cost of development and, perhaps, discouraging such development altogether.

DEVELOPMENT STANDARDS FOR NEW DEVELOPMENT

A comparison of the development standards from the 1986 report with those cited in the current survey suggests that little has changed. Table 5 provides a summary of the most common standards and indicates that, in most cases, the average or range of development parameters set by the communities that responded are identical to those cited in the 1986 report. Some of the responses indicated that, insofar as maximum density, lot width, and rear yard setback standards were concerned, there has been no change over the past decade. The discussion that follows takes a closer look at each of these standards.

Minimum Size of Development

At one time, zoning authorities believed that manufactured housing developments should have at least 10 acres and 50 home sites "in order to provide a full range of facilities and an economic return sufficient to support good management." The 1986 survey revealed a trend toward lower minimum parcel-size requirements; indeed, most of the communities in that study required fewer than 10 acres for a manufactured housing development or had no requirement. Results of the current survey suggest that most local governments continue to believe that a low parcel-size requirement is appropriate or that no minimum should be set. More than two-thirds of the 20 communities examined in this section of the report permit manufactured home developments smaller than 10 acres in size or have no requirement.

The 1986 report notes that the trend toward low or no minimum size requirement was a desirable one and was consistent with the general trend toward permitting smaller site-built developments, a trend that is still evident today, especially in urban redevelopment areas.

> Many communities now permit very small planned unit developments of only two to three acres and, in some cases, allow them as infill development in built-up areas. Establishing a low minimum parcel-size requirement for manufactured home developments can allow greater use of these factory-built units. Not having a minimum parcel-size requirement would provide even greater flexibility by allowing a case-by-case determination of what constitutes the optimal size of a manufactured home development. The latter approach takes into account market conditions and site characteristics that might limit what can be done with a given parcel of land.

Results of the current survey suggest that most local governments continue to believe that a low parcel-size requirement is appropriate or that no minimum should be set. More than two-thirds of the 20 communities examined in this section of the report permit manufactured home developments smaller than 10 acres in size or have no requirement.

Table 5. Comparison of Standards for Manufactured Housing Developments (1986 and 1996)

Description	1986	1996
Minimum size of development:		
No requirement	3 respondents	5 respondents
5 acres or less	10 respondents	9 respondents
10 acres	4 respondents	7
20 acres or more	3 respondents	2
Maximum density	6-8 units/acre	6-8 units/acre
Minimum lot area	3,500-5,000 sf	4,000-5,000 sf
Minimum lot width	40-50 ft.	40-50 ft.
Maximum lot coverage	50-65%	40-50%
Minimum setback:		
front	20 ft	20-25 ft
side	4-8 ft	10 ft
rear	10 ft	10 ft
Minimum unit separation	10-20 ft	10-20 ft
Minimum common open space	150-400 sf/unit or 5-10% of total area	100-300 sf/unit or 8-10% of total area
Perimeter requirements	20-50 ft buffer or 6-8 ft wall/fence	25-50 ft buffer or 6 ft wall/fence
Landscaping	required	required
Minimum street width	18-26 ft (access)	20-25 ft (access)
Sidewalk width	3-4 ft	2-4 ft
Minimum parking	2 spaces	2 spaces

The earlier report provides the following concluding remarks on the subject of minimum size of development that are worth restating here.

> If a community believes that a minimum size requirement must be set, it should consider a low minimum. A minimum parcel size of three to five acres...would appear to be adequate for...(land-lease or rental communities) A similar minimum parcel-size requirement for manufactured home subdivisions could also be set but seems unnecessary.

Only a few of the communities examined, in both 1986 and in the more recent survey, had minimum size requirements for manufactured home subdivisions, and such requirements are rarely imposed on subdivisions that contain site-built housing.

Density

About half of the communities featured in Table 6 have placed a cap on the density or number of units per acre permitted in manufactured housing developments. All but two of these communities have a density cap that falls within six to eight units per acre. Albany, Oregon, permits manufactured housing developments to be up to 10 units per acre and Poplar Bluff, Missouri, allows land-lease communities to be developed at nine units per acre. Poplar Bluff is also one of only two communities that place a cap on the density of manufactured home subdivisions at 8.7 units per acre; the other is Johnson County, Kansas, which requires that both land-lease and subdivision developments can not exceed five units per acre.

Among the communities that do not place a cap on density in their development provisions, two—West Sacramento, California, and Pinellas County, Florida—require that the density of these developments should be consistent with limits established in their comprehensive or general plan. Grand Island/Hall County, Nebraska, which uses an overlay zone to permit manufactured housing developments in existing zoning districts, requires that the density of such development must be the same as that established for the underlying district.

Densities in the range of six to eight units per acre, which are common in the communities featured in this report, are reasonable. In fact, eight units per acre is probably the absolute maximum density that can be achieved in a well-planned land-lease community. This is especially the case when a certain amount of the land area of the development is devoted to common open space. The lower end of the range (six units per acre) would appear to be an entirely appropriate limit for manufactured housing subdivisions, particularly if common open space is not provided. In most cases, the density standard for manufactured housing subdivisions is the same as the one established for site-built subdivisions. In either case, a density cap may not be necessary if a minimum lot-size requirement is set, especially when larger lots (say 4,500 square feet and greater) and some common open space are required. When small lots are allowed and a community wants to exercise strict control over density, a cap can and perhaps should be placed on density.

Lot Size

All but two of the 20 communities examined in this section and featured in Table 7 have set minimum lot-size requirements for manufactured housing developments. The two communities that have not established minimum requirements for manufactured housing developments require either that such development is consistent with site-built development (Albany, Oregon) or that minimum lot size must adhere to the standards set in the general plan (West Sacramento, California).

In most cases, the communities that have established a minimum lot size require that lots should be at least 4,000 square feet and, in some instances, considerably larger. Brookings, South Dakota, requires that manufactured housing developments have a minimum lot size of 7,000 square feet, and Johnson County, Kansas, requires that the lots in certain manufactured home subdivisions must be at least 10,000 square feet.

It would appear that the trend toward large minimum lot-size requirements for manufactured home developments, noted in the 1986 report, continues. Nearly three-fourths of the communities responding to the 1970 survey permitted lots that were 3,500 square feet or less in manufactured

Densities in the range of six to eight units per acre, which are common in the communities featured in this report, are reasonable. In fact, eight units per acre is probably the absolute maximum density that can be achieved in a well-planned land-lease community. This is especially the case when a certain amount of the land area of the development is devoted to common open space. The lower end of the range (six units per acre) would appear to be an entirely appropriate limit for manufactured housing subdivisions, particularly if common open space is not provided. . . . When small lots are allowed and a community wants to exercise strict control over density, a cap can and perhaps should be placed on density.

Table 6. Minimum Size and Maximum Density Requirements

Jurisdiction	Minimum Size of Development	Maximum Density (number of units per gross acre, unless otherwise specified)
Manufactured Home Developments:		
Albany, Oregon	5 acres	10
Hampton, Virginia	5 acres	No requirement
Inver Grove Heights, Minnesota	No requirement	No requirement
Poplar Bluff, Missouri	5 acres	*subdivision* RS-4: 7.2 per net acre RS-5: 8.7 per net acre *land-lease developments* RS-5: 9 per net acre
Portland, Maine	No requirement	No requirement
Roanoke County, Virginia	5 acres	*land-lease developments* 7 *subdivisions* No requirement
Scappoose, Oregon	2 acres	No requirement
West Sacramento, California	5 acres	Established in General Plan
Manufactured Home Development Districts:		
Brookings, South Dakota	No requirement	6
Burnsville, Minnesota	No requirement	No requirement
Charlotte Mecklenburg County, North Carolina	2 acres	6
Grand Island Hall County, Nebraska	Varies depending on underlying district	Varies depending on underlying district
Hays, Kansas		
land-lease developments	5 acres	7
subdivisions	10 acres	No requirement
Johnson County, Kansas		
land-lease developments	10 acres	5
subdivisions	25 acres	5
Lancaster, Texas	20 acres	No requirement
Owensboro-Daviess County, Kentucky	10 acres	8
Pinellas County, Florida		
subdivisions	10 acres	Density limits established in comprehensive plan
land-lease developments	No Requirement	8; less if maximum density permitted by comprehensive plan is less
Springfield, Missouri	10 acres	8
Wichita-Sedgwick County, Kansas	5 acres	No requirement
Winston Salem-Forsyth County, North Carolina	4 acres	5

Table 7. Minimum Lot Area, Width/Frontage, and Maximum Coverage Requirements

Jurisdiction	Lot Area	Width/frontage	Maximum Lot Coverage
Manufactured Home Developments:			
Albany, Oregon	Same standards apply as for conventional development.		
Hampton, Virginia			
land-lease developments			
single-section	3,800 sq. ft.	40 feet	No requirement
multisection	5,225 sq. ft.	55 feet	No requirement
subdivisions			
single-section	4,725 sq. ft.	45 feet	No requirement
multisection	6,300 sq. ft.	60 feet	No requirement
Inver Grove Heights, Minnesota	5,000 sq. ft.	50 feet	No requirement
Poplar Bluff, Missouri			
RS-4	6,000 sq. ft.	50 feet	50%
RS-5	5,000 sq. ft.	45 feet	50%
Portland, Maine	4,500 sq. ft.	50 feet	50%
Roanoke County, Virginia	4,000 sq. ft.	40 feet	No requirement
Scappoose, Oregon	2,500 sq. ft.	No requirement	No requirement
West Sacramento, California	"Density. . .shall not exceed density range as defined in the general plan for the property on which the park is located."		

housing land-lease communities. The 1986 survey found that about half of the 20 communities singled out for study had minimum lot sizes ranging from 4,000 to 5,000 square feet. Two-thirds of the communities that responded to the 1996 survey have set minimum lot-size requirements that exceed 4,000 square feet, with most of those specifying lot sizes of 4,000 to 5,000 square feet. Its important to note, however, that, while minimum lot-size requirements may have increased over the years, these minimums are still well within the range considered appropriate to such development. More importantly, it is necessary to keep lot sizes in this range to allow for the development of affordable developments, whether the homes are built on site or in a factory.

Three of the communities that have set minimum lot-size requirements—Winston Salem-Forsyth County, North Carolina, Hampton, Virginia, and Johnson County, Kansas—permit smaller lots when single-section homes are used. They require larger lots for multisection homes. Hampton, Virginia, for example, allows lots as small as 3,800 square feet for single-section homes in manufactured home land-lease communities but has set a 5,225-square-foot minimum lot size for multisection homes.

Some of the communities featured here also require that manufactured housing subdivisions provide greater lot area than land-lease communities. For instance, Pinellas County, Florida, which permits lots as small as 3,500 square feet in land-lease communities requires a minimum lot size nearly twice that large (6,000 square feet) in manufactured subdivision developments.

Two-thirds of the communities that responded to the 1996 survey have set minimum lot-size requirements that exceed 4,000 square feet, with most of those specifying lot sizes of 4,000 to 5,000 square feet. Its important to note, however, that, while minimum lot-size requirements may have increased over the years, these minimums are still well within the range considered appropriate to such development.

Jurisdiction	Lot Area	Width/frontage	Maximum Lot Coverage
Manufactured Home Development Districts:			
Brookings, South Dakota	7,500 sq. ft.	50 feet	No requirement
Burnsville, Minnesota	3,600 sq. ft. (interior) 4,500 sq. ft. (corner)	40 ft. (interior) 50 ft. (corner)	No requirement No requirement
Charlotte Mecklenburg County, North Carolina	5,000 sq. ft.	40 ft.	No requirement
Grand Island/Hall County, Nebraska	Same as underlying zone but not less than 6,000 sq. ft.	Same as underlying zone	Same as underlying zone
Hays, Kansas			
land-lease developments	3,600 sq. ft.	40 ft.	No requirement
subdivisions	6,000 sq. ft.	50 ft.	No requirement
Johnson County, Kansas			
land-lease developments			
single-section	4,000 sq. ft.	45 ft.	No requirement
multisection	5,000 sq. ft.	55 ft.	No requirement
subdivisions			
single-section	7,500 sq. ft.	75 ft.	No requirement
multisection	10,000 sq. ft.	85 ft.	No requirement
Lancaster, Texas			
individual lots (fee simple)	6,600 sq. ft.	60 ft.	40%
land-lease developments	4,500 sq. ft.	40 ft.	No requirement
Owensboro-Daviess County, Kentucky			
land-lease developments	2,000 sq. ft.	No requirement	33.3%
subdivision	3,000 sq. ft.	No requirement	33.3%
Pinellas County, Florida			
land-lease developments	3,500 sq. ft.	20 ft.	40%
subdivisions	6,000 sq. ft.	60 ft.	40%
Springfield, Missouri	4,000 sq. ft.	40 ft.	40%
Wichita-Sedgwick County, Kansas	5,000 sq. ft.	40 ft.	No requirement
Winston Salem/ Forsyth County, North Carolina			
single-section	4,000 sq. ft.	40 ft.	No requirement
multisection	5,000 sq. ft.	50 ft.	No requirement

Poplar Bluff, Missouri, varies its minimum lot-size requirement in relation to the district in which the development will occur. In this city's RS-4 residential district, the minimum lot-size requirement for manufactured housing developments is 6,000 square feet; in the less restrictive RS-5 residential district, the minimum lot size permitted is 5,000 square feet.

Allowing for some flexibility in minimum lot-size requirements recognizes that manufactured dwellings vary significantly in size and that single-section homes will require less lot area than multisection homes. Similarly, larger lots are often more appropriate in subdivision development since these developments are more likely to be permitted in existing residential districts where larger site-built homes and larger lots are more common.

The range of minimum lot-size requirements featured in this section of the report are reasonable and, more importantly, they should not have serious negative effects on the affordability of manufactured housing. All but four of the 18 communities that have established minimum lot-size requirements have standards that call for lots of 3,500 to 5,000 square feet. Lots smaller than 3,500 square feet could be permitted, especially when single-section homes are used and depending on how other development parameters are handled, including setback, yard, frontage, and common open space requirments. When smaller lots are permitted, communities must make sure that these and other standards do not result in developments in which lots are crowded, units are sited too close together, and common open space is lacking. Scappoose, Oregon allows lots to be as small as 2,500 square feet in land-lease developments, but, when lots will be this small, it also requires that 100 square feet of open space must be provided for each home for a "recreational play area, group or community activities." This play or related activity area dedication is not required if the individual lots in the development are 4,000 square feet or larger.

A minimum lot size ranging anywhere from 3,500 to 5,000 square feet is entirely reasonable for multisection units. In most cases a 4,000-square-foot lot can accommodate the typical multisection home with no problem, and a smaller minimum could work, depending on other development parameters.

Lot Width or Frontage

Three-fourths of the communities featured in Table 7 have established minimum lot-width requirements, ranging from 20 feet in Pinellas County, Florida, where the minimum lot size for manufactured home land-lease communities is 3,500 feet to 85 feet in Johnson County, Kansas, where the minimum lot size in a manufactured housing subdivision with multisection homes is 10,000 square feet. In most cases, however, minimum lot-width requirements are between 40 and 50 feet, which is the same range cited in the 1986 report. (See Table 5 on page 29.) Requiring 40 to 50 feet of lot frontage is reasonable when the minimum lot size is 4,000 to 5,000 square feet. When smaller lots are permitted, less lot width may be required.

The amount of lot frontage or width required or actually needed varies in relation to how the unit is sited and other development standards, including building setback requirements, parking, and, as noted above, minimum lot size. When units can be sited parallel to the side lot line or when only one side yard is required, as is the case in zero lot line development, lot frontage may be kept to a minimum. Less frontage is also needed when space for parking or a garage does not have to be placed at the front of the lot. From the standpoint of affordability, it is especially important that frontage requirements are not excessive. The length of streets, sidewalks, and utility runs are determined by the amount of lot frontage or width.

Lot Coverage

About a third of the communities featured in this section have set maximum lot-coverage requirements to help prevent the crowding of lots and to ensure that a portion of the lot is devoted to open space. (See Table 7.) Owensboro-Davies County, Kentucky, permits only 33 percent of the lot in a manufactured housing development to be covered by buildings. It should be noted, however, that this community also permits lots that are only 2,000

Lots smaller than 3,500 square feet could be permitted, especially when single-section homes are used and depending on how other development parameters are handled, including setback, yard, frontage, and common open space requirments. When smaller lots are permitted, communities must make sure that these and other standards do not result in developments in which lots are crowded, units are sited too close together, and common open space is lacking.

The amount of lot frontage or width required or actually needed varies in relation to how the unit is sited and other development standards, including building setback requirements, parking, and, as noted above, minimum lot size. When units can be sited parallel to the side lot line or when only one side yard is required, as is the case in zero lot line development, lot frontage may be kept to a minimum. Less frontage is also needed when space for parking or a garage does not have to be placed at the front of the lot. From the standpoint of affordability, it is especially important that frontage requirements are not excessive. The length of streets, sidewalks, and utility runs are determined by the amount of lot frontage or width.

square feet—the smallest minimum lot size among the communities featured here. In the other communities that have established a maximum coverage requirement, lot coverage standards range between 40 to 50 percent. These maximum allowances for lot coverage are reasonable for the 4,000- to 5,000-square-foot lots required in most cases. Permitting less coverage would seem appropriate for the 2,000-square-foot lots permitted in Owensboro-Davies County, Kentucky.

Unit Setback and Separation

All but three of the 20 communities featured in Table 8 have established minimum front, side, and rear setback requirements. Front setbacks

Jurisdiction	Front Setback (feet)	Side Setback (feet)	Rear Setback (feet)	Unit Separation (feet)
Manufactured Home Developments:				
Albany, Oregon	8	3	3	10
Hampton, Virginia *land-lease developments*	15	10, from one lot line 15, from the other	10	No requirement
subdivisions	20	15	15	No requirement
Inver Grove Heights, Minnesota	18	No requirement	15	20
		"Corner lot shall have 18-foot setbacks from the curb line to any structure."		
Poplar Bluff, Missouri				
RS-4	15	6	10 if unit is perpendicular to fronting street	No requirement
			20 if unit is parallel to fronting street	
RS-5	15	6	10 if unit is perpendicular to fronting street	20 for and-lease developments
			20 if unit is parallel to the fronting street	
Portland, Maine	20	10	20	No requirement
Roanoke County, Virginia	20	5	10	26
Scappoose, Oregon	20	10	20	10
		"...on a corner lot the street side yards shall be 20 feet."		
West Sacramento, California	No requirement	No requirement	No requirement	15

Table 8. Minimum Unit-Setback and Separation Requirements (continued)

Jurisdiction	Front Setback (feet)	Side Setback (feet)	Rear Setback (feet)	Unit Separation (feet)
Manufactured Home Development Districts:				
Brookings, South Dakota	20	6	25	No requirement
Burnsville, Minnesota	15	10 15 (street)	10	No requirement
Charlotte Mecklenburg County, North Carolina				
land-lease development	20	10	10	No requirement
subdivision	"In a subdivision, the lot and yards should be developed to the standards of the R-5 (residential) district."			
Grand Island/Hall County, Nebraska	Same as underlying zone	10	Same as underlying zone	No requirement
Hays, Kansas				
land-lease development	10	7.5	10	20
subdivision	25	5	25, or 20 percent of the depth of the lot, whichever is smaller	No requirement
Johnson County, Kansas	25	10	10	30
Lancaster, Texas				
individual lot (fee simple)	30	5 20 (side street)	10	No requirement
land-lease development	15 (private street) 30 (public street)	5	10	20
Owensboro-Daviess County, Kentucky	10	5	5	10
Pinellas County, Florida				
interior lots	10	5	5	No requirement
perimeter lots	25	10	10	No requirement
Springfield, Missouri	25	5	20 percent of the lot depth, but not less than 10 feet nor more than 25 feet	15
Wichita-Sedgwick County, Kansas	10	10	10	10
Winston Salem/ Forsyth County, North Carolina	20	5, with a combined width of 15 feet	10	No requirement

ensure that the dwelling unit is not sited too close to the street, and side or rear setbacks allow a certain amount of separation between units and create yard area. Inver Grove Heights, Minnesota, has established requirements for only front and rear setbacks; Grand Island/Hall County, Nebraska, has established a minimum standard only for side yard setbacks; and West Sacramento, California, has not established minimum requirements for any area of the lot but has established a minimum unit-separation requirement.

Twelve of the 20 communities have established unit-separation requirements that, in a number of cases, apply only to land-lease communities. (See Table 8.) In all but two communities, these standards call for 10 to 20 feet between homes or between homes and other structures. These development parameters appear to be reasonable but may be altogether unnecessary if unit-setback requirements are properly drawn. Most communities, however, prefer to use both sets of controls to ensure that lots in manufactured housing development will not be crowded and that homes will also not be sited too close to other buildings (e.g., a community center) in the development.

Most of the communities that have established minimum front setback standards require that the unit be located 15 to 20 feet from the front lot line; 20 feet of front yard setback is the most common standard. This amount of front setback is often required in small-lot development, especially when two parking spaces or a garage must be accommodated at the front of the lot. Less setback may be needed if parking can be placed at the rear of the lot or off the lot altogether. At least 20 feet of front setback may also be needed on a narrow street in order to facilitate backing a car off the lot and into the street, but, when more street width is provided, less front setback is required. Clearly, a case can be made for flexible setback requirements; some communities, for instance, vary the amount of setback required in small-lot site-built developments, depending upon how parking will be handled or how homes are sited on the lots.

Almost all of the communities have established setback requirements for side and rear years in manufactured housing developments. Most communities require 5 to 10 feet of side yard setback and 10 feet of rear yard setback. These development parameters are reasonable. It should be noted, however, that a number of communities now allow one side yard to be eliminated altogether to allow for zero lot line siting of homes. The remaining side-yard must usually be at least 10 feet.

Common Open Space and Facilities

Nearly three-fourths of the communities featured in Table 9 require that manufactured housing developments contain common open space. Most communities require that a certain percentage of the net or total area of the development must be reserved for common open space. The remaining communities require a certain amount of open space for each dwelling unit. The percentage of open space, based on either the net or total area of the development ranged from 5 percent of the net developable land area of a subdivision located in the RS-4 district of Poplar Bluff, Missouri, and 5 percent of the total land area in Inver Grove Heights, Minnesota, up to 20 percent of the total area of the development, as required by Springfield, Missouri.

The amount of open space required per dwelling unit ranged from 100 square feet by both Winston-Salem Forsyth /County, North Carolina, and Scappoose, Oregon, up to 300 square feet per lot in developments with 20 or more lots, in Hampton, Virginia. Winston-Salem/Forsyth County actually requires the greater of 4,000 square feet of open space or 100 square feet per

Clearly, a case can be made for flexible setback requirements; some communities, for instance, vary the amount of setback required in small-lot site-built developments, depending upon how parking will be handled or how homes are sited on the lots.

Table 9. Minimum Common Open Space and Perimeter Requirements

Jurisdiction	Required Common Open Space	Perimeter Requirements
Manufactured Housing Developments:		
Albany, Oregon	200 sq. ft. per home of outdoor or indoor recreation area that may be in one or more locations, at least one area must have minimum dimensions of 50 ft. by 100 ft. A separate play area for children under 14 years; the area must be at least 2,500 sq. ft. with at least 100 sq. ft. of area per home. Separate play areas are not required if development is either restricted to children over 14 or if home lots are at least 4,000 sq. ft.	Buffering and screening is required.
Hampton, Virginia	Land-lease and subdivision developments with 20 or more lots must provide 300 sq. ft. of "green area" per lot to be retained as common area. This area must be aggregated in increments of at least 4,500 sq. ft. This area must be landscaped, and no more than 25% of it may consist of water area.	Land-lease and subdivision developments must "provide a perimeter screen composed of landscaping or a combination of landscaping and fencing, the intent of which is to limit ingress and egress on the property, and to provide some buffering from adjoining uses. The screen shall be at least six feet in height and shall be located on all project property lines that do not abut existing or proposed public rights-of-way; such screening shall be set back at least 10 feet from any existing or proposed public right-of-way." "No structure, except fences as part of the perimeter screen, shall be permitted within 50 feet of the project property lines. Parking shall not be located closer than 20 feet from the property lines. All green area and landscaping within this buffer, except for that required for the perimeter screen, may be counted toward the green area [common open space] requirement."
Inver Grove Heights, Minnesota	5% of the land area of the development.	A 50-foot buffer must be provided in those parts of the development that are adjacent to single-family residential areas. A 20-foot buffer is required where the development is adjacent to a public street, residential development, commercial or industrial area, or park land. The buffer must be planted with a dense combination of trees, shrubs, and bushes as to form a screen to adjoining properties.

Jurisdiction	Required Common Open Space	Perimeter Requirements
Manufactured Housing Developments:		
Poplar Bluff, Missouri	Subdivisions in RS-4 districts must provide no less than 5% of the net developable land area for common open space. Subdivisions and land-lease developments in the RS-5 districts must provide no less than 10% of the net developable land area for common open space. The minimum size of a single parcel of ground for common open space for a subdivision or land-lease development must not be less than 7,500 sq. ft.	"Buffering and/or screening shall be required along the exterior boundaries of manufactured home developments which adjoin residential areas. Screening or buffering may be required in other locations when a nuisance or obnoxious use would interfere with the enjoyment of the proposed development."
Portland, Maine	No requirement	"The entire development shall be properly screened from abutting neighborhoods and uses. Such screen shall consist of plantings, or a combination of earth berm and plantings, not less than three feet in width and six feet in height at the time of initial occupancy of such development. . . . Individual shrubs or trees, as approved by the city arborist, shall be planted so as to establish a dense visual screen year round. At least 50% of the plantings shall consist of evergreens."
Roanoke County, Virginia	8% of the gross area of the development; the minimum "countable" space must be 5,000 contiguous sq. ft. "shall include passive and active facilities and be of an appropriate nature and location to serve the residents . . . may include facilities, such as recreation centers, swimming pools, tennis and basketball courts, and similar facilities."	"A Type C buffer yard. . .shall be installed along the side and rear perimeter of the . . . (development)."
Scappoose, Oregon	100 sq. ft. for each home must be provided for "a recreational play area, group, or community activities, but no recreational area shall be less than 2,500 sq. ft. No recreation area is required if the individual manufactured home spaces contain 4,000 sq. ft. or more."	No requirement
West Sacramento, California	No requirement	A land-lease development must be "enclosed by a masonry wall of at least seven feet in height located on property side of the street landscape setback. . .and along all property lines adjoining another private property."

Jurisdiction	Required Common Open Space	Perimeter Requirements
Manufactured Home Development Districts:		
Brookings, South Dakota	Standards established for individual lots, no specific development provisoins for manufactured home development.	
Burnsville, Minnesota	10% of the land area of the development must be developed for recreational use.	"A bufferyard of not less than 30 feet in width shall be landscaped with appropriate grass, shrubbery, and trees around the entire perimeter of the development."
Charlotte-Mecklenburg County, North Carolina	At least 8% of total area of the development must be devoted to recreational use by residents. "Such use may include space for community buildings, gardens, outdoor play area, ball courts, racquet courts, etc."	"No structure shall be located within 30 feet of any property line defining the perimeter [of the development]"
Grand Island/Hall County, Nebraska	No data.	No data.
Hays, Kansas	200 sq. ft. per home space; individual recreational areas must not be less than 5,000 sq. ft.; at least 50% of the recreational facilities must be constructed prior to the development of 50% of the project and must be completed by the time the project is 75% developed.	"A solid or semi-solid fence or wall, minimum six feet, maximum eight feet high, shall be provided between the manufactured home district and any adjoining property or property immediately across the alley which is zoned for residential purposes other than. . .manufactured homes." "In lieu of said fence or wall, a landscape buffer may be provided not less than 25 feet in width and said landscape buffer shall be planted with coniferous and deciduous plant material so as to provide proper screening. . . . When the landscape buffer is used . . . [it] shall not be included as any part of a required rear yard."
Johnson County, Kansas	"At least one private area shall be provided. . . . The size of such recreation area(s) shall not be less than 10% of the gross area of the . . . [development]."	"Effective screening shall be provided along boundaries of any. . .[development] adjoining industrial, commercial, or lower density residential uses or zoning districts to serve as a buffer through the use of plantings, fencing, berms, or other landscaping features. At a minimum, the perimeter. . .shall be planted with shade and ornamental trees to accent and help visually screen the development." Landscaping: Shade and ornamental trees should be planted on the perimeter of the development and also provided within a land-lease development at a ratio of one tree for each three lots and wherever practical. Entrances must be accented with plantings of shrubs, ornamental trees, or shade trees. Such landscaping within the development and at the entrances is encouraged for manufactured housing subdivisions.

Jurisdiction	Required Common Open Space	Perimeter Requirements
Manufactured Home Development Districts:		
Lancaster, Texas	No requirement	No requirement
Owensboro-Daviess County, Kentucky	"Common areas for recreational, management or service facilities should be of adequate area and configuration to accommodate contemplated structures and uses, and should be conveniently located to all residents."	Required (No data)
Pinellas County, Florida	*Land-lease development:* 10% of the gross site area must be devoted to open space and recreation facilities. *Subdivisions:* No requirement	No requirement No requirement
Springfield, Missouri	*Land-lease development:* 20% of the total area of the manufactured housing development, including required yards and bufferyards. Open space can not include areas covered by buildings, parking area, driveways, and internal streets. Open space must contain living ground cover and other landscaping materials. *Subdivision:* 30% of total lot area must be devoted to open space, including required yards and and bufferyards. Open space can not include areas covered by buildings, parking areas, driveways, and internal streets. Open space must contain living ground cover and other landscaping materials.	Minimum yard along boundaries: 25 feet. "Whenever any development in an R-MHC district is located adjacent to a different district or a nonresidential use in an R-MHC district is located adjacent to a residential use in an R-MHC district, screening and a bufferyard shall be provided. . . ." "Whenever any development in an R-MHC district is located adjacent to a different zoning district or a nonresidential use in an district is located adjacent to a residential use in an R-MHC district, screening and a bufferyard shall be provided. . . ."
Wichita-Sedgwick County, Kansas	No requirement	*Land-lease development:* "All structures within. . .[the development] shall be setback at least 20 feet from public street rights-of-way. . . ." *Subdivisions:* "All structures within . . . [the subdivision] shall be setback at least 10 feet from public street rights-of-way. . . ."
Winston Salem/Forsyth County, North Carolina	4,000 sq. ft. or 100 sq. ft. per manufactured home, whichever is greater.	A bufferyard of a minimum width of 30 feet must be established along each exterior property line except, where adjacent to a private street or public right-of-way external to the development, the bufferyard must be 50 feet.

dwelling unit. Scappoose also requires that the open space must be at least 2,500 square feet, but open space is not required if lots in the development are 4,000 square feet or larger.

Hampton, Virginia, requires that common areas must be aggregated in increments of at least 4,500 square feet and be landscaped; no more than 25 percent of the common area may consist of water area. Popular Bluff, Missouri, requires a considerably larger common area of 7,500 square feet for both subdivisions and land-lease developments. The city does, however, permit payments in lieu of land dedication that can be used to develop recreational facilities.

Owensboro-Daviess County, Kentucky, requires common areas and facilities in manufactured housing developments but has not established specific requirements. Rather, the county allows the specifics to be determined on a case-by-case basis. The following provisions are in the county's zoning law.

> **Common Areas and Facilities.** Common areas for recreational, management, or service facilities should be of adequate area and configuration to accommodate contemplated structures and uses, and should be conveniently located to all residents.

Most of the communities that require open space include improvement standards in their zoning provisions to help ensure that the space is usable. Roanoke County, Virginia, for example, requires that common area must include facilities allowing for both passive (e.g., sitting) and active (e.g., game playing) activities "and be of an appropriate nature and location to serve residents. . . . This may include facilities such as recreation centers, swimming pools, tennis and basketball courts, and similar facilities."

To help ensure that recreational facilities are built, Hays, Kansas, requires that the construction of these facilities must be completed by the time the manufactured housing community is 75 percent developed:

> Each manufactured home [development] shall devote a minimum of 200 square feet per manufactured home space for recreational area. Individual recreational areas shall not be less than 5,000 square feet and required setbacks, roadways, and off-street parking spaces shall not be considered as recreation space. A minimum of 50 percent of the recreational facilities shall be constructed prior to the development of one-half of the project, and all recreational facilities shall be constructed by the time the project is 75 percent developed.

These provisions further stipulate that recreational facilities may be located in a community building that also includes "laundry facilities, storm shelter, and other similar uses."

Some amount of common open space in residential development is usually desirable, and this is especially the case when small lots are permitted. The amount provided, however, may be less important than whether or not it is improved and usable. The open space requirement might be waived or modified if the development is located near existing public open space.

Perimeter Requirements

All but four of the communities featured in Table 9 have established perimeter requirements for manufactured housing developments. In most cases, dwelling units, other structures, or lots must be set back from the boundaries of the development to create a buffer area between dwelling units and structures, and adjacent land uses. This buffer area

---

Most of the communities that require open space include improvement standards in their zoning provisions to help ensure that the space is usable.

Some amount of common open space in residential development is usually desirable, and this is especially the case when small lots are permitted. The amount provided, however, may be less important than whether or not it is improved and usable. The open space requirement might be waived or modified if the development is located near existing public open space.

usually cannot be used to satisfy the open space requirement and must be landscaped. Inver Grove Heights, Minnesota, Hampton, Virginia, and Portland, Maine, also require that this area include plantings or fencing that can serve to screen the development from abutting uses. Poplar Bluff, Missouri, is the only community that requires buffering and screening in manufactured housing developments, but it does not include specific requirements in its zoning provisions for these features, presumably allowing these issues to be decided on a case-by-case basis. The Poplar Bluff ordinance requires only that:

> Buffering and/or screening shall be required along the exterior boundaries of manufactured home developments which join residential areas. Screening or buffering may be required in other locations when a nuisance or obnoxious use would interfere with the enjoyment of the proposed development.

Specific perimeter standards for manufactured housing developments vary. Some communities that responded to the survey require a 10-foot buffer; others require 50 feet of setback from the development's boundary to provide an adequate buffer area. Most communities, however, require between 25 and 50 feet of perimeter setback. In some cases, this standard will vary based on the nature of adjacent uses. In Johnson County, Kansas, for example, more buffer area is required along a public right-of-way:

> **Perimeter yard requirement.** No part of any manufactured home or other building or structure shall be located within 50 feet of any public road right-of-way nor within 25 feet of any exterior property line of the manufactured home [land-lease development] or manufactured home subdivision.

Inver Grove Heights, Minnesota, on the other hand, requires more buffer area when the exterior boundary of the development is adjacent to a single-family residential area.

> A 50-foot buffer shall be provided. . .in those locations where the [manufactured home land-lease development] is adjacent to single-family residential areas. A 20-foot buffer shall be required in each [manufactured home land-lease development] where the [development] is adjacent to a public street, multifamily residential development, commercial or industrial [uses], or park land.

Poplar Bluff, Missouri, requires buffering or screening only along the exterior boundary of a manufactured home development that is adjacent to a residential area. In Hays, Kansas, manufactured housing developments may use either a "solid or semisolid fence or wall" when they are adjacent to residential uses other than manufactured homes, or "in lieu of said fence or wall, a landscape buffer may be provided not less than 25 feet in width and said landscape buffer shall be planted with coniferous and deciduous plant material so as to provide proper screening for the [development]."

Flexible standards that allow for a case-by-case determination of the specific amount or type of buffer required or whether any buffer is necessary are generally preferable to more specific perimeter requirements. Standards that offer an option of fencing or setback are also more desirable than those that require either a buffer or fencing or both. The appearance and quality of units in the development, the type of adjacent nearby uses, and the quality of the development plan can all have an effect on how much perimeter setback is necessary and whether a fence or landscaping should be required.

Specific perimeter standards for manufactured housing developments vary. Some communities that responded to the survey require a 10-foot buffer; others require 50 feet of setback from the development's boundary to provide an adequate buffer area. Most communities, however, require between 25 and 50 feet of perimeter setback. In some cases, this standard will vary based on the nature of adjacent uses.

Standards that offer an option of fencing or setback are also more desirable than those that require either a buffer or fencing or both. The appearance and quality of units in the development, the type of adjacent nearby uses, and the quality of the development plan can all have an effect on how much perimeter setback is necessary and whether a fence or landscaping should be required.

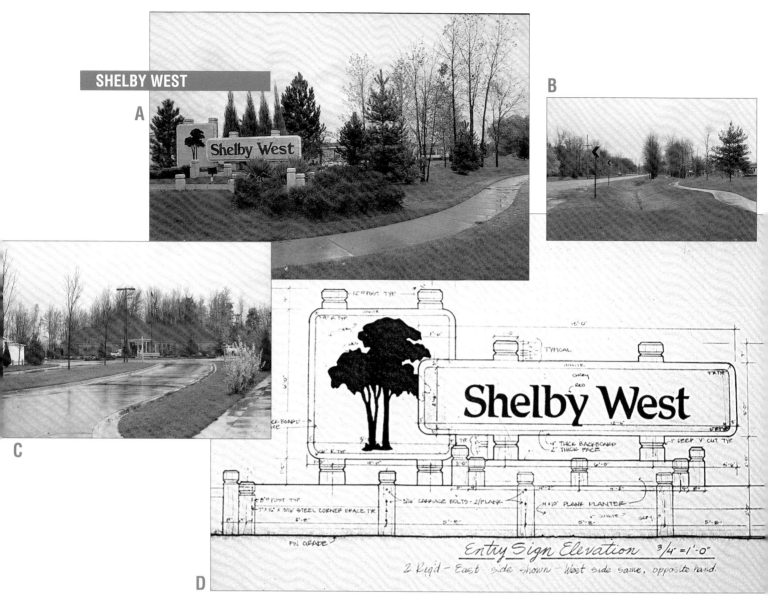

SHELBY WEST

A

B

C

D

Shelby West

Entry Sign Elevation 3/4" = 1'-0"
2 Rigid - East side shown - West side same, opposite hand.

Many communities require that manufactured home developments include a common area and a landscaped buffer at the perimeter to help ensure compatibility with and screening from adjacent land uses. The Shelby West development in the Detroit metropolitan area offers some good examples. The entrance (A), perimeter (B) and common area (C) of this development are extensively landscaped. Common area improvements include a wetlands preserve and community center. The aesthetics of this development are further enhanced by an "image" package that includes a carefully designed signage scheme (D).

Landscaping

In addition to requiring that common open space and perimeter buffer areas be landscaped, many communities have landscaping provisions for other areas of manufactured housing developments. Johnson County, Kansas, requires landscaping within the manufactured housing development and at the entrance(s) of the development.

> The perimeter of the of the manufactured home development shall be planted with shade and ornamental trees to screen and accent the development as required. . . . Shade and ornamental trees shall also be provided within manufactured home [land-lease developments] at a ratio of one tree for each three lots and shall be located wherever practicable within the development. Entrances to manufactured home [land-lease developments] shall be accented with plantings of shrubs, ornamental trees, or shade trees. Such landscaping within the development and at the entrance(s) is encouraged for manufactured home subdivisions.

Portland, Maine, requires trees in manufactured housing developments for the expressed purpose of enhancing the appearance of the manufactured units.

> Each unit shall be provided with at least two trees meeting the city's arboricultural specifications and which are clearly visible from the street line and are located so as to visually widen the narrow dimensions or proportion of the unit.

Landscaping is especially important in small-lot developments like manufactured housing land-lease communities and subdivisions. In addition to its aesthetic purpose, landscaping can be used to enhance privacy, to help conserve energy by shading and cooling the buildings and the roadway surface, and to give some visual order to the development. And, as noted above, Portland, Maine, requires landscaping to "visually widen the narrow dimension or proportion" of a manufactured unit. Local regulations for manufactured housing development should require or at the vary least encourage the use of landscaping.

Streets and Sidewalks

In most cases, standards required that streets be between 20 and 24 feet wide for local access or minor streets where no on-street parking is allowed. (See Table 10.) The most common minimum requirement for street width was 20 feet. When parking was allowed on one side of the street, the minimum pavement width allowed was between 27 to 30 feet, and, when parking was allowed on both sides of the street, a minimum width of 32 to 40 feet was allowed. One community, Inver Grove Heights, Minnesota, varies its minimum street-width requirement in relation to expected traffic patterns and density.

> Minimum width between curb faces shall be 32 feet for collector street and 28 feet for minor streets, unless projected traffic patterns and density shall determine a greater width, at which time the greater width shall be determined by the City Engineer.

Ideally, street width should be determined on a case-by-case basis, taking into account expected average daily traffic, amount of lot frontage, whether on-street parking will be permitted, how parking will be handled on the lot, and other factors related to street design. In most cases, 18 to 20 feet of pavement, allowing for two moving lanes of traffic but no parking on the street, should be adequate for both site-built and manufactured home developments. Each travel lane would be at least nine feet wide and parking lanes should be at least eight feet wide.

The communities featured in Table 10 that have established specific sidewalk or walkway standards for manufactured housing developments require that they must be between two to three feet in width. Some communities, including Johnson County, Kansas, Hays, Kansas, and Charlotte- Mecklenburg, North Carolina, require that manufactured housing developments include walkways or sidewalks, but do not establish specific minimum width requirements. Johnson County, Kansas, for example requires that "an all-weather surfaced walk system shall be provided for pedestrian traffic along at least one side of all streets in the development and along streets adjacent to the development."

In Inver Grove Heights, Minnesota, the planning commission can grant a variance if it finds that a sidewalk is not necessary.

> A concrete sidewalk, not less than 30 inches wide, shall be constructed adjacent to a street. [For] lots located within areas, such as cul-de-sacs, where in the opinion of the planning commission, sidewalks shall serve no useful purpose, a variance can be granted. The variance shall require the Council's approval.

Ideally, street width should be determined on a case-by-case basis, taking into account expected average daily traffic, amount of lot frontage, whether on-street parking will be permitted, how parking will be handled on the lot, and other factors related to street design.

Table 10. Minimum Street-Width and Sidewalk Requirements

Jurisdiction	Width of Street Pavement	Sidewalk/Walkway Requirements
Manufactured Home Developments:		
Albany, Oregon	24-foot accessway if no parking allowed; 30 feet if parking is allowed on one side, and 36 feet if parking is allowed on both sides. First 50 feet of the accessway measured from the public street must be surfaced to a minimum width of 30 feet.	"Permanent walkways of not less than three feet in width shall be provided from each manufactured home main entrance to the nearest public or private street. A minimum of four-foot-wide walkways shall connect each manufactured home space with common areas, public streets, and play areas. All walkways must be separated, raised, or protected from vehicular traffic and provide access for handicapped persons."
Hampton, Virginia	22 feet, if no parking 27 feet, if parking on one side of street 32 feet, if parking on both sides of street	No requirement
Inver Grove Heights, Minnesota	20 feet, if one-way street 28 feet, if minor street 32 feet, if collector "unless projected traffic patterns and density shall determine a greater width . . . the greater width will be determined by the city engineer."	"A concrete sidewalk, not less than 30 inches wide shall be constructed adjacent to the street. [For] lots located within areas, such as cul-de-sacs, where, in the opinion of the planning commission, sidewalks shall serve no useful purpose, a variance can be granted."

Poplar Bluff, Missouri	"The development shall provide publicly dedicated streets. Required improvements shall be provided in compliance with the street paving standards described in the [code]. Looped and cul-de-sac streets may be provided" as follows:
	    40-foot right-of-way
	    20-foot paving widths
	    "when these streets serve less than 20 dwelling units and and when off-street guest parking is provided."
		No requirement

Portland, Maine	No data	No data
Roanoke County, Virginia	11 feet, if one-way minor, no parking 20 feet, if minor, no parking 28 feet, if minor, parking on one side 30 feet, if collector, no parking 36 feet, if collector, parking both sides	"Manufactured home lots not served by a public or private street may be served by a walkway, trail, or bikeway, provided that such pathway serves the front, rear, or side of the manufactured home lot. Each pathway shall be constructed of a hard-surface or gravel material and shall have a minimum width of three feet."
Scappoose, Oregon	20 feet, if no on-street parking 30 feet, if on-street parking	"Walkways shall connect each manufactured home to its driveway. All walks must be concrete, well drained, and not less than 35 inches in width."
West Sacramento, California	25 feet, if no parking on street 35 feet, if parking on one side of street 45 feet, if parking on both sides of street	No requirement

Table 10. Minimum Street-Width and Sidewalk Requirements (continued)

Jurisdiction	Width of Street Pavement	Sidewalk/Walkway Requirements
Manufactured Home Development Districts:		
Brookings, South Dakota	No data	No data
Burnsville, Minnesota	No data	No data
Charlotte Mecklenburg County, North Carolina	"Internal streets and circulation patterns shall be adequate to handle the traffic to be generated by the development."	"A walkway shall be constructed for each lot or space to connect parking spaces to the manufactured home entrance."
Grand Island/Hall County, Nebraska	No data	No data

Hays, Kansas	Standards for private roadways:
	    24 feet, if no on-street parking
	    29 feet, if parking on one side of street
	    40 feet, if parking on both sides of street
		"Common walks shall be provided in locations where pedestrian traffic is concentrated. . . . Common walks should preferably be through interior areas removed from the vicinity of streets."

Johnson County, Kansas	"All internal streets shall comply with the Street Construction Standards [for subdivision development] adopted by the County."	"The provision and maintenance of an all-weather surfaced walk system shall be provided for pedestrian traffic along at least one side of all streets in the development and along streets adjacent to the development.

Owensboro-Daviess County, Kentucky	*Serving up to 40 lots:*
	    22 feet, if local access, no parking
	    27 feet, if local access, on-street parking
	Serving more than 40 lots:
	    24 feet, if local subcollector, no parking
	    34 feet, if local subcollector, with parking
	Serving more than 90 lots:
	    26, if minor collector street, no parking
	    37, if minor collector street, with parking
		"All manufactured home [lots] shall be connected to common walks or to streets, or to driveways, or to parking spaces. Such individual walks shall be a minimum width of two feet."

Pinellas County, Florida	*Land-lease development:* May be private and built to the following dimensions: No data.
	    minor street: 20 feet (paved surface); 25 feet (right-of-way)
	    collector street: 24 feet (paved surface); 30 feet (right-of-way)
	    subdivisions: No data	

Springfield, Missouri	No data	No data
Wichita-Sedgwick County, Kansas	No data	No data
Winston Salem-Forsyth County, North Carolina	"Each manufactured home space shall have direct vehicular access to an internal private access easement and street."	"A hard surface walkway, being a minimum of two feet wide, leading from the entrance of the manufactured home to its parking spaces or to the street shall be constructed."

In most cases, a three-foot minimum sidewalk width is adequate. The width of sidewalks or walkways that provide direct access to individual homes could be reduced to two feet, and three- or four-foot sidewalks could be used to provide access to common areas or facilities. Albany, Oregon, varies the width of sidewalks in manufactured home developments.

> Permanent walkways of not less than three feet in width shall be provided from each manufactured home main entrance to the nearest public or private street. A minimum of four-foot-wide walkways shall connect each manufactured home space with common areas, public streets, and play areas. All walkways must be separated, raised, or protected from vehicular traffic and provide access for. . .[disabled] persons.

In many cases, sidewalks may be needed only on one side of the street and do not have to follow alongside the street.

Parking

Most of the communities examined in this section and featured in Table 11 require a minimum of two parking spaces per dwelling unit, which, in many cases, have to be located on the lot they serve. Some communities, including Roanoke County, Virginia, Hampton, Virginia, and Springfield, Missouri, allow at least one of the required parking spaces to be located off the lot it serves. Springfield, Missouri, allows both required spaces to be located off the lot provided that "one of the two spaces required shall be within 100 feet of the lot served."

Some communities require parking for guests. Johnson County, Kansas, requires that manufactured housing land-lease communities must provide one guest parking space for every three manufactured units in the development. Johnson County also requires that all planned manufactured home land-lease developments must have an area(s) set aside for "the storage of boats, trailers, automobiles, and other equipment for seasonal or periodic use." The county further stipulates that:

> Such equipment shall not be stored upon a manufactured home lot nor upon the streets within the manufactured home [land-lease development]. Such storage areas are encouraged in manufactured home subdivisions.

Allowing some flexibility in the placement of parking on or off the lot is a good idea. Lots in some manufactured housing developments may be too small to adequately accommodate parking spaces. At least two parking spaces per unit is, in most cases, reasonable, and provisions should be made for guest parking. Manufactured home developments designed for an older population, however, may have less need for parking than other developments, and, in such cases, local governments may want to require less parking, both for residents and guests.

STANDARDS FOR EXISTING DEVELOPMENT

Growing concern over the condition of some older manufactured housing developments is prompting a number of communities to consider what can be done to improve the appearance and overall livability of these communities.

Some communities are including provisions in their ordinances that address this problem. West Sacramento, California, mandates that certain improvements be made when there is an "enlargement or extension" of an existing development. In Winston-Salem/Forsyth County, North Carolina, all existing manufactured housing developments must meet certain standards within four years. The county has also established

Allowing some flexibility in the placement of parking on or off the lot is a good idea. Lots in some manufactured housing developments may be too small to adequately accommodate parking spaces. At least two parking spaces per unit is, in most cases, reasonable, and provisions should be made for guest parking. Manufactured home developments designed for an older population, however, may have less need for parking than other developments, and, in such cases, local governments may want to require less parking, both for residents and guests.

Table 11. Parking Requirements

Jurisdiction*	Minimum Number of Spaces Required
Manufactured Home Developments:	
Albany, Oregon	two per home and one guest space for every eight units
Hampton, Virginia	two off-street parking spaces for each home.
	"At least one space shall be provided on the lot housing the unit. The additional space may be provided in an off-street parking area . . . located within 150 feet of the unit it is to serve."
Inver Grove Heights, Minnesota	two off-street spaces per home
Poplar Bluff, Missouri	two spaces to serve each dwelling unit
Roanoke County, Virginia	"Each manufactured home lot shall have the equivalent of two parking spaces. At least one of these spaces shall be provided on the manufactured home lot, unless the lot is accessed by a pathway. . . ."
Scappoose, Oregon	two spaces per home.
West Sacramento, California	"There shall be the equivalent of two parking spaces per manufactured home site."
Manufactured Housing Development Districts:	
Burnsville, Minnesota	two off-street parking spaces per unit. Manufactured home developments "shall maintain a hard surface off-street parking lot for guests of occupants of at least one space for each five [homes]."
Johnson County, Kansas	"Each manufactured home lot shall have off-street parking space for at least two automobiles. . . . One parking space for every three manufactured home lots must be provided for guests."
Lancaster, Texas	"Two covered enclosed parking spaces per unit behind the front yard line shall be required."
Owensboro/Daviess County, Kentucky	Two off-street spaces per home, located on the lots they serve.
Springfield, Missouri	
land-lease developments	Two per home
	"Parking spaces for each manufactured home do not have to be provided on each lot, however one of the two required parking spaces required shall be located within 100 feet of the lot served."
subdivision	No data

*Other communities that had responded to the survey did not list parking requirements. Rather than list them here as having "no data," we have excluded them.

provisions for the expansion of developments that are not in compliance with these standards.

Existing Manufactured Housing Developments

1. *Schedule for Improvements.* Manufactured housing developments lawfully existing at the time of the adoption of this ordinance shall be required to meet the following standards of this section within four years of the ordinance's adoption date:

(a) *Bufferyards*. A type II bufferyard of minimum width of 30 feet shall be established along each exterior property line, except where adjacent to a private street or public right-of-way not internal to the development. Along external private streets or public right-or-way, a type II bufferyard of a minimum of 50 feet shall be established, with the exception of meeting minimum width requirements in developments where meeting the width requirements would result in the relocation of structures or manufactured homes.

(b) *Solid Waste*. Each space shall have a minimum of one solid waste container with a tight fitting cover and a capacity of not less than 32 gallons, or dumpsters of adequate capacity may be substituted. If dumpsters are provided, each such container shall be located on a concrete slab and screened on three sides by an opaque fence at least eight feet in height.

(c) *Skirting*. Each manufactured home shall have skirting installed in accordance with the following requirements:
   (1) Skirting shall be of noncombustible material that will not support combustion. Skirting material shall be durable and suitable for exterior exposures.
   (2) Any wood framing used to support the skirting shall be of approved moisture-resistant, treated wood.
   (3) The skirting shall be vented in accordance with state requirements.
   (4) Skirting manufactured for this purpose shall be installed in accordance with the manufacturer's specifications.
   (5) Skirting shall be properly maintained.

(d) *Utilities*
   (1) Location. All utilities within a manufactured home development shall be located underground.
   (2) Water. Connection to a public water system and installation of fire hydrants meeting the standards of the appropriate jurisdiction are required.
   (3) Sewer. Connection to a public sewer system or installation of an approved package treatment plant is required. . .unless public water and sewer is located more than 200 feet from the manufactured home development.

(e) *Streets*. Streets shall have a minimum of four inches of gravel and be well maintained.

2. *Expansion of Nonconforming Manufactured Housing Developments*. No expansion of a nonconforming housing development shall be permitted unless all units in the development, both preexisting and additional, have vertical skirting or a similar structural enclosure around the entire base of the unit between the outer walls and the ground or paved surface, and are anchored to the ground in accordance with the regulations set forth by the State of North Carolina for manufactured and modular housing units.

These standards can help to improve the appearance of older manufactured housing communities. At some point, however, older homes will need to be replaced with new homes that are usually larger than the older ones they replace. This will require reworking the site plan of the older development and reconfiguring lots and streets. An alternative is provided by a Newport Beach, California, development that is featured in Part 4 of this report. Older homes in that development are being replaced with a narrow two-story home that can be placed on existing lots. This innovation and others are discussed in Part 4.

Part 4

Design Innovations

Once relegated to manufactured housing developments only, the single-section home is now used in infill development. Garages and porches are now being used to make these homes more compatible with existing homes in infill situations, and these units are being designed with two levels.

This section looks at how the design and appearance of manufactured housing and manufactured housing developments have evolved over the past decade, considers current trends, and offers some observations on what can be expected in the years to come. A decade ago, most new manufactured homes were single-section homes, and most of these smaller homes were placed in land-lease developments. Only about a quarter of the manufactured units produced at that time were multisection homes. Today, larger, multisection homes represent more than half of the new units being manufactured annually, and a growing number of these larger homes are designed to be indistinguishable from site-built housing. The appearance of some single-section homes has also improved. Many of the newer manufactured housing developments include amenities that were once offered only in upscale, site-built developments. More than ever before, manufactured housing subdivisions are being developed. A look at several projects will attest to the significant changes that have taken place in both individual units and manufactured housing developments.

THE IMPROVING APPEARANCE OF THE SINGLE-SECTION HOME

The demand for the single-section manufactured housing unit has declined in recent years. Still, the appearance of these units has improved. The most affordable dwelling unit built today, the single-section manufactured home has been undergoing a series of changes to its appearance that make it more acceptable. Some signs of this were evident more than a decade ago with the use of single-section units in planned unit developments, including the HUD Affordable Housing Demonstration project in Elkhart County, Indiana, in the early 1980s and the HomePlus development in Pomona, California, that was developed in the mid-1980s. The single-section home has grown wider, increasing from an average of 14 feet in width a decade ago to 16 feet and, in some states, 18 feet. Once relegated to manufactured housing developments only, the single-section home is now used in infill development. Garages and porches are now being used to make these homes more compatible with existing homes in infill situations, and these units are being designed with two levels.

Urban Infill Development

More than a decade ago, Architect Paul Wang and Associates developed Laurel Courts, their first manufactured home development in Oakland, California. This 30-unit infill development included a mix of single-section and multisection homes with attached site-built, one-car garages. A zero lot line siting configuration was employed and the appearance of each home was enhanced by the addition of courtyards, trellises, and landscaping.

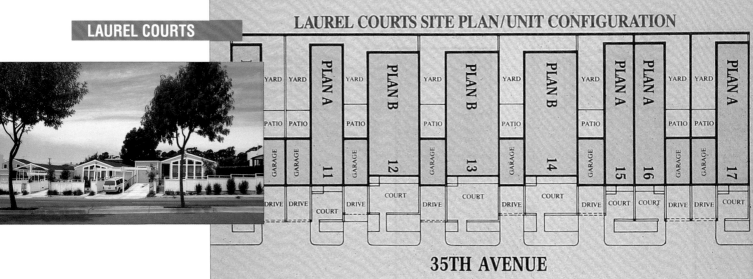

LAUREL COURTS

LAUREL COURTS SITE PLAN/UNIT CONFIGURATION

YARD	YARD	PLAN A	YARD	PLAN B	YARD	PLAN B	YARD	PLAN B	YARD	PLAN A	PLAN A	YARD	YARD	PLAN A
PATIO	PATIO		PATIO		PATIO		PATIO		PATIO			PATIO	PATIO	
GARAGE	GARAGE	11	GARAGE	12	GARAGE	13	GARAGE	14	GARAGE	15	16	GARAGE	GARAGE	17
DRIVE	DRIVE	COURT	DRIVE	COURT	DRIVE	COURT	DRIVE	COURT	DRIVE	COURT	COURT	DRIVE	DRIVE	COURT

35TH AVENUE

Each home in the Laurel Courts development in Oakland, California, includes a windowless side wall to allow for zero lot line siting of homes on narrow urban infill lots. The zero lot line siting configuration helps to create private outdoor space on each lot. The appearance of single-section homes was enhanced by the addition of garages, courtyards, trellises, and landscaping.

Paul Wang's current development consist of single-section and multisection manufactured units that are placed on infill lots in the Elmhurst neighborhood in Oakland. Elmhurst is an older inner-city neighborhood that has a number of vacant lots and deteriorated housing that need to be replaced. The Wang project is an effort to help increase property values and prevent further decline in the area. The new single-section homes are selling for about $90,000; the typical cost of a new site-built home in the area is $170,000. The single-section homes are being placed on narrow infill lots (25 feet or less), and the appearance of these homes is enhanced by the addition of a site-built garage, porch, and pitched (ratio of 4:12) roof.

Paul Wang submitted a new infill development proposal to the City of Oakland, California, in 1997 that will place four or five homes on two separate blocks and two homes on an adjacent block. The 11 single-section homes proposed in the Martin Luther King Jr. Plaza development include eight different floor plans and five different elevations. Each home will include a site-built garage, porch, and entryway. Like the Laurel Courts development, manufactured units are being used to achieve affordability.

ELMHURST

This manufactured home fits in well with existing homes in this area of Oakland. The appearance of this single-section home is enhanced by site-built additions, including the entryway and fencing.

Single-section homes with garages are being placed on small, narrow lots in this older inner-city neighborhood in Oakland.

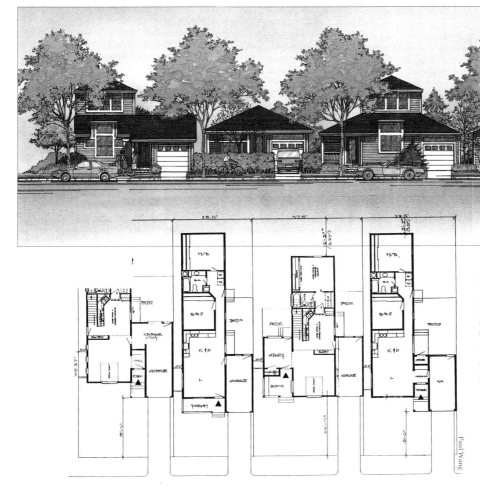

MARTIN LUTHER KING, JR., PLAZA

Each single-section home in this infill development will include a site-built garage, porch, and entryway. Three of the homes planned for Allen Street will consist of stacked sections to create two-story homes.

REVITALIZATION OF A MANUFACTURED HOUSING COMMUNITY

It was noted above that, when older manufactured home communities or mobile home parks are renovated and old homes are replaced with larger new homes, the site plan of the older development will need to be reworked. Streets and lots may need to be reconfigured. The Riverbrook development in New Haven, Michigan, for example, needed to be reconfigured (see site plans on page 55) to accommodate new multisection homes that replaced single-section units. An alternative approach is exemplified by the Lido Peninsula development in Newport Beach, California, in which two-story homes were specially designed so that they could be placed on existing lots. The Lido Peninsula manufactured home community first opened in the 1950s as a mobile home park. Since 1996, the owner of this 214-lot community has encouraged tenants to replace their aging mobile homes with specially designed two-story manufactured homes when their leases come up for renewal. A number of the older homes have been replaced with the narrow two-story homes that fit the compact 35' x 30' lots in this development. Priced at $80,000, these new units are a real bargain in this California community that overlooks considerably more expensive homes and boats that dock at the waterfront nearby.

LIDO PENINSULA

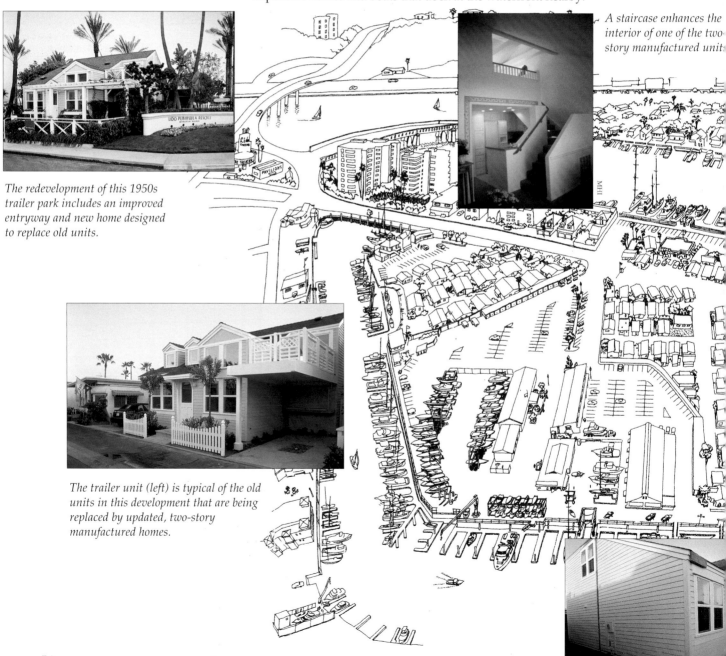

The redevelopment of this 1950s trailer park includes an improved entryway and new home designed to replace old units.

The trailer unit (left) is typical of the old units in this development that are being replaced by updated, two-story manufactured homes.

A staircase enhances the interior of one of the two-story manufactured units.

A

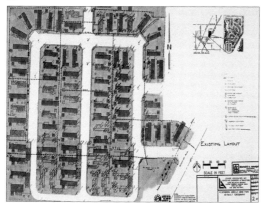

B

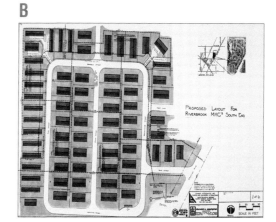

Nearly all the homes in the Riverbrook manufactured home community in New Haven, Michigan, are single-section units (A). The reconfigured site plan (B) calls for 75 percent of the homes to be multisection units. The reconfigured development has fewer units but is more responsive to current market demand.

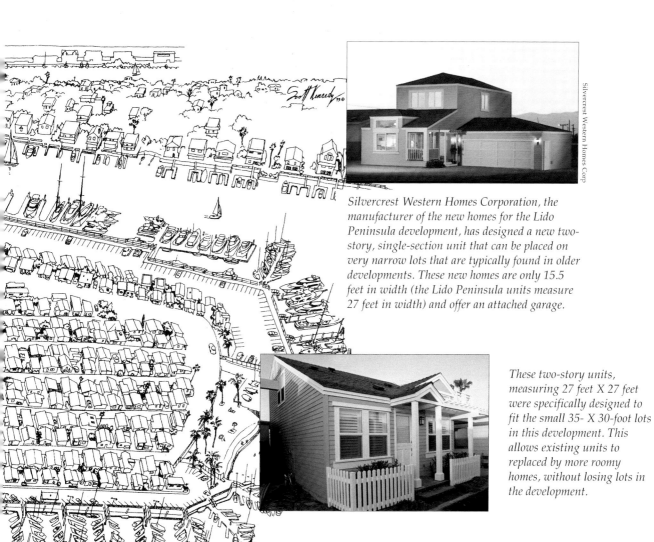

Silvercrest Western Homes Corporation, the manufacturer of the new homes for the Lido Peninsula development, has designed a new two-story, single-section unit that can be placed on very narrow lots that are typically found in older developments. These new homes are only 15.5 feet in width (the Lido Peninsula units measure 27 feet in width) and offer an attached garage.

These two-story units, measuring 27 feet X 27 feet were specifically designed to fit the small 35- X 30-foot lots in this development. This allows existing units to replaced by more roomy homes, without losing lots in the development.

homes are built and shipped to the
building site in two sections, which are
then joined at the site. Hardi-board fiber
cement siding is used so that the homes
will withstand the corrosive effects of
the salty and humid air of the area.

TWO-STORY UNITS

The single most important design innovation and refinement in manufactured housing over the past decade has been the two-story manufactured unit. A less dramatic but important occurrence is the placement of manufactured units over basements. The first large-scale development of two-story manufactured homes is the New Colony Village development in the Baltimore area. This 52-acre land-lease development will consist of 416 homes—350 two-story and 66 one-story homes—sited on 30 acres of land. This gated community includes 22 acres of wetlands and forest preservation land, a day care center, a convenience store, and a club house. Homes in this development range in size from a 940-square-foot, two-bedroom, one-story home up to a 1,540-square-foot, three-bedroom, two-story home. All of the homes have a garage; some have a two-car garage. Prices range from $97,990 to $130,990—not high for new single-family homes in Howard County, Maryland.

Factory-built sections are placed on a site-built first floor that includes a two-car garage. The finished home is pictured below.

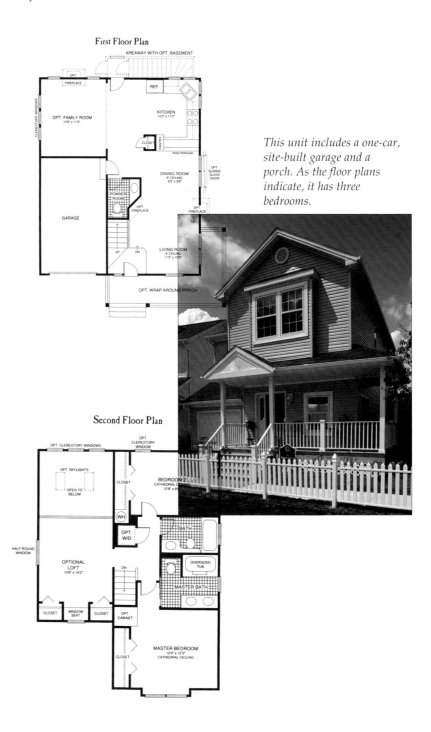

First Floor Plan

Second Floor Plan

This unit includes a one-car, site-built garage and a porch. As the floor plans indicate, it has three bedrooms.

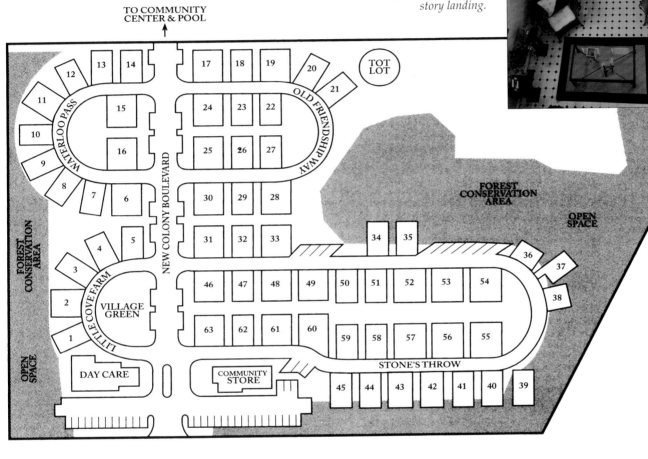

The plan (below) for the first phase of
New Colony Village includes a day care
facility, a community center and store,
and an ample amount of open space.
an interior shot (right) from a second-
story landing.

TO COMMUNITY
CENTER & POOL

TOT LOT

WATERLOO PASS

OLD FRIENDSHIP WAY

13 14
12
11 15
10 16
9
8 7 6

17 18 19 20
24 23 22 21
25 26 27
30 29 28
31 32 33

NEW COLONY BOULEVARD

FOREST CONSERVATION AREA

OPEN SPACE

34 35
36
37
38

46 47 48 49 50 51 52 53 54
63 62 61 60 59 58 57 56 55

LITTLE COVE FARM

VILLAGE GREEN

4 5
3
2
1

FOREST CONSERVATION AREA

OPEN SPACE

DAY CARE

COMMUNITY STORE

STONE'S THROW

45 44 43 42 41 40 39

The New Colony
Village streetscape
seen from a second-
story porch.

MHI

MANUFACTURED HOME SUBDIVISIONS

In recent years, manufactured housing subdivisions have become more common, and, with an increasing number of local governments permitting manufactured units on individual, fee simple lots, this trend should continue. In addition, a growing number of companies that develop site-built subdivisions and that have never worked with manufactured housing before are now involved in manufactured housing development. In Seattle, Washington, for example, HomeSight, a local community development corporation, is developing a subdivision that will include fee simple, detached and attached manufactured homes to achieve greater affordability.

In 1996, Centex Corporation, the nation's second largest builder of site-built homes, acquired a major manufactured housing company, and Cincinnati-based Zaring Homes, one of the largest Midwest housing developers, is now using manufactured homes in their subdivisions. Pulte Home Corporation, the nation's largest builder of site-built housing, created a manufactured housing division in 1996 called Canterbury Communities, which developed its first manufactured housing subdivision in the Raleigh, North Carolina, area. The first phase of this subdivision includes 57 homes that are priced from $90,000 to $125,000, which is about $30,000 less than comparably sized homes in nearby subdivisions. This development includes a mix of single-story manufactured units and two-story modular units.

LEXINGTON COMMUNITIES

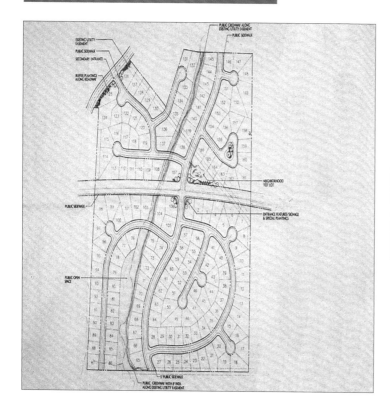

Lexington Communities is Pulte Homes Corporation's first manufactured home subdivision development. The plan for this development creates fee simple lots that average 9,000 square feet; improved common open space that includes a tot lot and public greenway with an eight-foot path; and provisions for the perimeter and entry areas of the development.

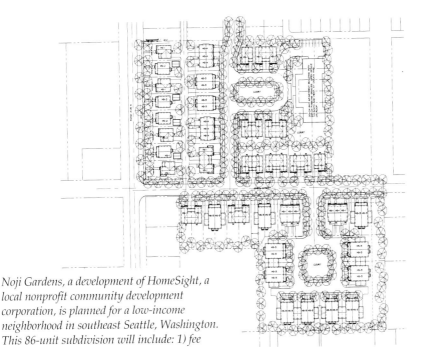

A

There are two single-family prototypes for the Noji Gardens project. The first (A) is based on one of the New Colony village homes in Baltimore, Maryland (see above). The second (B) is based on a prototype developed by Susan Maxman Partners, Ltd., for the Milwaukee Urban Design Project (see below). This unit includes about 1,400 square feet of space. The price for either of these units will be about $155,000; the median price of a single-family home in Seattle is $210,000.

B

Noji Gardens, a development of HomeSight, a local nonprofit community development corporation, is planned for a low-income neighborhood in southeast Seattle, Washington. This 86-unit subdivision will include: 1) fee simple, single-family detached homes with garages acceessible from an alley (see drawing along 32nd Ave. S); 2) duplex and fourplex manufactured homes (along Juneau Street and the surrounding courtyards); 3) fee simple triplex rowhouse units that may be site built, modular, or manufactured (at bottom of site plan); and a site-built condominium apartment building with 20 units. The condominium units will range in price from $85,000 to $100,000. This traditional site plan includes short blocks to encourage walking and an alley is provided to allow driveways and garages to be placed at the rear of some lots. Two centrally located and landscaped courtyards with tot lots and benches will also be provided. The developer will offer home ownership counseling, and down payment assistance will be available to some low- and moderate-income households.

C

D

Two different styles of fee-simple duplex units (C and D) will be offered in Noji Gardens. Like the single-family models depicted above (A and B), these units were inspired by the models in Baltimore and Milwaukee.

E

The fourplex unit (E) is created by attaching duplexes at their garages. These units will be priced at about $125,000.

LAND-LEASE COMMUNITIES

The design of manufactured homes in land-lease communities has also improved significantly since the 1986 PAS Report. Most notable has been the "developer series" homes that are designed to be virtually indistinguishable from site-built housing. These homes have conventional siding and steeper roofs that are constructed of traditional roofing materials. On-site additions such as garages, porches, gables, and special entryway treatments further enhance the appearance of these manufactured units.

Santiago Estates, developed in 1990 by Watts Industries, is one a number of manufactured home communities in California developed by a traditional developer whose previous work was limited to site-built housing. This Los Angeles, California, community is targeted to first-time buyers and is one of the first land-lease communities in which manufactured homes can be purchased as real property through conventional 30-year, fixed-rate mortgages. Homes range in size from 1,250 to 1,625 square feet and were priced from $94,900 to $113,900, with a monthly lease payment of about $466. The average cost of a comparable site-built home in the Los Angeles area in the early 1990s was $250,000. The homes in this development include central air conditioning, two-car garages, and landscaping with sprinklers.

DEVELOPER SERIES

A

B

C

D

E

F

Three examples of "developer series" manufactured homes in California, designed to be virtually indistinguishable from site-built housing; Canyon View Estates development in the City of Santa Clarita in Los Angeles County was developed by Canyon View Partnership and includes 400 homes priced in 1990 from $79,000 to $101,000 with two- and three-bedroom, two-bath units that range in size from 1,100 to 1,600 square feet. Stucco exteriors and tile roofs gives homes in the Rancho Viejo development in Escondido an appearance that is consistent with regional housing styles. Homes in Santiago Estates (C), Los Angeles, California, range in size from 1,250 to 1,625 square feet. Prices range from $94,900 to $113,900. These "developer series" homes have many enhancements (D, E, F) that add to their attractiveness.

URBAN INFILL

Many low- and moderate-income, inner-city families spend more than half their income on housing that is often substandard and poorly maintained. In many older urban areas, rehabilitating dilapidated housing has been the primary means of increasing housing opportunities for these families. Increasingly, this approach is proving to be too costly, especially when the objective is to resell these rehabilitated units to needy households. A growing number of local officials are considering the use of manufactured housing as an alternative to more costly new site-built and rehabilitated housing.

In the past, manufactured units have rarely been used in urban infill development. Local zoning ordinances may exclude such housing from existing residential neighborhoods where affordable new housing is badly needed, and, in many cases, the traditional manufactured housing styles may not be entirely compatible with existing housing styles in older urban neighborhoods. In 1995, the MHI launched the Urban Design Project to demonstrate that manufactured housing can be designed so that it is compatible with existing housing in established urban neighborhoods and yet be more affordable.

MHI hired Susan Maxman and Partners, Ltd., a nationally recognized architectural firm to design manufactured homes for urban infill lots in five cities—Wilkinsburg, Pennsylvania; Birmingham, Alabama; Washington, D.C.; Louisville, Kentucky; and Milwaukee, Wisconsin. In each city, the architects and MHI staff have worked with local residents, developers and lenders, public officials, and manufacturers to develop manufactured housing designs that are appropriate for urban infill. These efforts are highlighted in the following pages. The Milwaukee initiative, which is part of an ongoing neighborhood planning effort and builds on an earlier MHI-sponsored Design Studio, is examined in greater detail on pages 65-68.

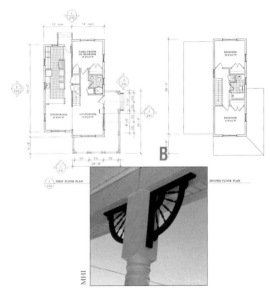

A

B

C

WILKINSBURG

Wilkinsburg, Pennsylvania, a community adjacent to Pittsburgh, was chosen as the pilot location to kick-off the Urban Design Project. The Wilkinsburg project plans to build and place four manufactured homes on scattered infill lots. The first, a two-story, three-bedroom home, was placed on a corner lot at Kelly Street and Mifflin Avenue in 1997. The project development team in Wilkinsburg, coordinated here and in the other communities by Susan Maxman and Partners, Ltd., and MHI staff, included the Pennsylvania Department of Community and Economic Development, Wilkinsburg officials, ACTION Housing, Inc. (a local nonprofit housing developer), and New Era Building Systems, Inc. (the manufacturer). Residents of the Kelly Street and Mifflin Avenue area were involved in the development process through a series of focus groups. The information gained from the focus group sessions was especially useful to the project architects in their efforts to design the first home. Decorative porches (A) and post detailing (B) ensures that the manufactured housing designed for infill lots in Wilkinsburg resembles existing housing in the area. Hardi-board, a durable fiber cement material (C), provides a rich, woodlike finish that can be painted.

61

BIRMINGHAM

In Birmingham, Alabama, two home designs—a two-bedroom (A) and a three-bedroom (B)—have been developed through a series of meetings with the Smithfield Neighborhood Association. Each design is a two-section, single-story home that resembles the bungalow-style homes that are found in the neighborhood. The development team included the City of Birmingham's Office of Enterprise Communities and Planning Department, Alabama Power, the Alabama Manufactured Housing Institute, and Cavalier Homes. The first home will be built in 1998.

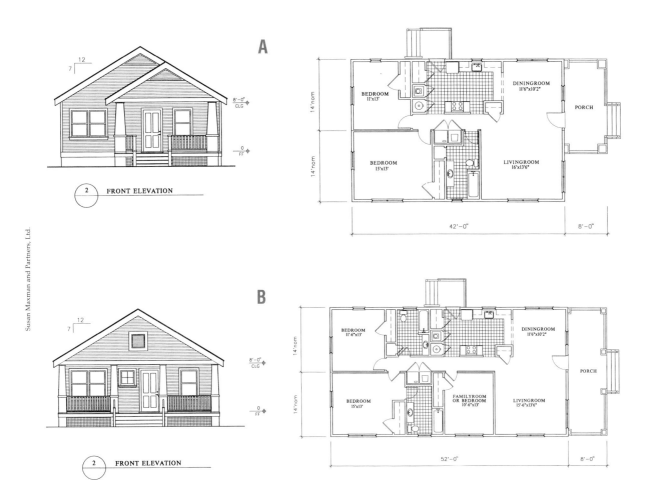

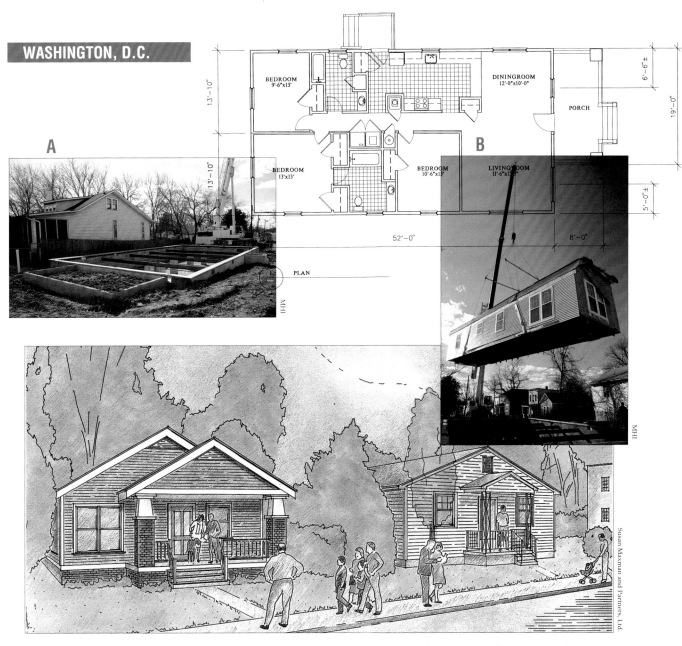

A

B

PLAN

BEDROOM
9'-6"x13'

DININGROOM
12'-0"x10'-0"

PORCH

BEDROOM
13'x13'

BEDROOM
10'-6"x13'

LIVINGROOM
11'-6"x13'

13'-10"

13'-10"

52'-0"

8'-0"

Susan Maxman and Partners, Ltd.

The Marshall Heights Community Development Organization, with the support of the Potomac Electric Power Company, plans to build two manufactured homes, including a three-bedroom, single-story home that will be built in two sections, and a two-story home. The one-story unit was completed in the spring of 1998. The manufacturer for the Washington project is Schult Homes Corporation. The floor plan for the one-story unit is shown above. The unit is placed on a foundation (A) by crane (B) and finishing work (C and D) is done on site, resulting in the completed home (E).

C

D

E

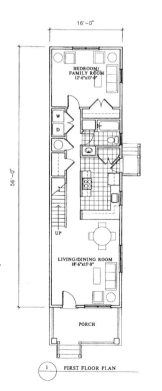

IHW

Susan Maxman and Partners, Ltd.

The Louisville, Kentucky, project consists of six infill sites in the Smoketown neighborhood, an area of the city that is undergoing a major redevelopment effort. A stacked single-section home has been placed on one of the lots selected, and single-section homes will be placed on the five other sites. The two-story home employs a "camel back" design that sets back the second level so that there is only a single story at the front lot line. This style of housing became common in Louisville, dictated by a now-defunct local property tax scheme that taxed property based on the number of stories at the front lot line.

Susan Maxman and Partners, Ltd.

1 FRONT ELEVATION

The preference among residents that took part in focus group sessions was that the design of new infill homes should embrace the "camel back" design. The stacked, single-section configuration also allows for a manufactured home to be placed on a very narrow, urban infill lot that measures only 21 feet in width. The development team includes the Neighborhood Development Corporation (a local nonprofit group), the Louisville Economic Opportunity Corporation, and New Era Building Systems, Inc. (the manufacturer).

1 FIRST FLOOR PLAN

2 SECOND FLOOR PLAN

Efforts in Milwaukee to design affordable manufactured housing for inner-city neighborhoods began in 1995 with a Design Studio sponsored by MHI and conducted by the University of Wisconsin-Milwaukee School of Architecture and Urban Planning. This graduate-level studio resulted in several prototypes for inner-city lots and was a source of ideas for MHI's Urban Design Project. The studio also served to introduce city housing development officials, community groups, state housing finance officials, and the general public to manufactured housing. Equally important, the positive media attention that this studio got helped to create a receptive environment for manufactured housing in the city.

Paul Olsen

A

MILWAUKEE DESIGN STUDIO

Ground Floor

Scott Starks

These prototypes demonstrate how manufactured housing could be designed so that it would be compatible with existing housing in Milwaukee's Midtown neighborhood. These prototypes also helped city and state housing officials and the general public to better understand manufactured housing. They can be taken apart and reassembled to illustrate the efficiencies of homes built in a factory and the manner in which each factory-built module is joined and/or stacked at the building site. Two of the nine models designed in the studio were two-section, 1.5-story homes (A); the other designs were two-story homes (B). The two sections of the 1.5-story model shown here are offset to create porches at the front and rear of the unit and the roof is hinged to permit a steep pitch that is typical of other homes in this older inner-city neighborhood.

Paul Olsen

B

The design schemes developed in the studio were based on these existing homes in the Midtown neighborhood.

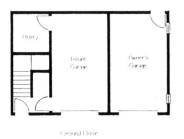

Ground Floor

Second Floor

Karen Mierow

The Urban Design Project in Milwaukee is a collaborative effort, involving local and state officials, including the Wisconsin Manufactured Housing Association and the state's top housing finance agency, the Wisconsin Housing and Economic Development Authority (WHEDA). At the local level, the city of Milwaukee and community-based organizations played key roles in the project. The project was officially kicked-off with a meeting in December 1996 involving the project's architects, MHI personnel, the mayor, and the city's Director of Housing and Community Development. Available city-owned lots in the Midtown Triangle neighborhood, which was selected as the location for new homes, were identified at that time.

There were several reasons why the Midtown Triangle neighborhood was chosen: new affordable housing is badly needed in this inner-city area; there is an abundance of vacant lots and, in some areas, largely vacant blocks; and a traditional neighborhood plan, sponsored by the city, had been developed and was being implemented. The plan for this 63-block area, close to downtown, called for a mix of affordable housing options at various income levels. A number of Habitat homes for low-income families have been built in this planning area, and the city developed the 43-unit, middle-income CityHomes subdivision near the blocks planned for manufactured homes. The CityHomes development has been very successful—all 43 homes had been sold within two years of breaking ground. The manufactured homes planned for this area were to be priced for low- and moderate-income households.

The Community Development Corporation of Wisconsin, an established nonprofit development group experienced in both rehabilitation and new construction, was selected as project developer, and Schult Homes of Indiana was chosen to manufacture the homes that Susan Maxman and Partners, Ltd., would design.

A key element of the design process was a series of meetings early in the development process. These meetings involved residents of the Midtown Triangle neighborhood, city and state officials, neighborhood organizations, lenders, Realtors, and other interested parties. The information gathered from these meetings was used to design several different prototypes.

The redevelopment strategy that the Milwaukee Urban Design Project is pursuing involves identifying blocks that are substantially vacant. Then several new for-sale homes will be sited on a block. The belief is that this approach will have a greater impact on the redevelopment of the area than a scattered siting approach. The first block selected for new homes includes more than 10 vacant lots; only two or three lots are occupied. The plan is to construct up to eight homes in the first phase of this project and then market other units. What types of units will make up the second phase of the project will be determined by how successful the sale of the first-phase units goes. The homes chosen for the first phase of the project include two, two-story, three-bedroom, and one and one-half bath designs. Each unit consists of four modules; one home totals 1,464 square feet of floor area and the other is 1,536 square feet.

Support for construction of the new homes includes Community Development Block Grant (CDBG) funds provided by the city of Milwaukee and a grant from Fannie Mae. Low-interest mortgages and grants for certain closing costs are available from WHEDA. Construction of the first homes in the Midtown Triangle neighborhood began in July 1998.

The Urban Design Project has generated a good deal of interest in manufactured housing and has prompted a number of local housing producers to consider the use of manufactured housing in proposed developments. The project has also led to some serious discussion of the possibility of locating a housing production plant in the city that could respond to a host of needs, including jobs, meaningful training, neighborhood revitalization, and affordable housing.

CITYHOMES

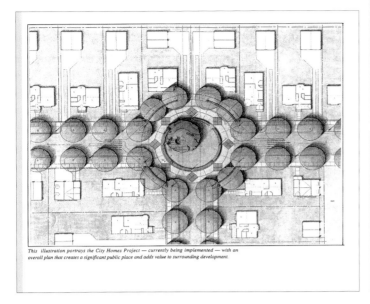

This illustration portrays the City Homes Project — currently being implemented — with an overall plan that creates a significant public place and adds value to surrounding development.

The design of homes in the recently completed CityHomes subdivision, located within two blocks of the planned manufactured home development, had to be considered by the project's design team. Like the CityHomes unit, the manufactured homes must also be two-story homes and include porches, a raised foundation, and basement. This subdivision was one of a number of new housing initiatives called for in the Midtown Triangle Neighborhood Plan.

MILWAUKEE DESIGN PROJECT

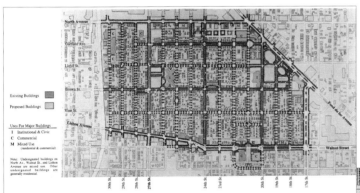

An abundant number of vacant lots and in some cases entire vacant blocks (see map below), mostly owned by the city of Milwaukee, are available for new homes in the development area. The map to the left shows the proposed redevelopment scheme as per the neighborhood plan

...lapidated units can be replaced with manufactured ...mes. The Midtown neighborhood has experienced ...siderable decline over the past 20 years. The ...uation offers many opportunities for siting new ...nufactured homes—provided they are compatible ...th existing homes on adjacent or nearby lots.

67

A sketch of the street view from one of the proposed manufactured homes in the Midtown Triangle Neighborhood development.

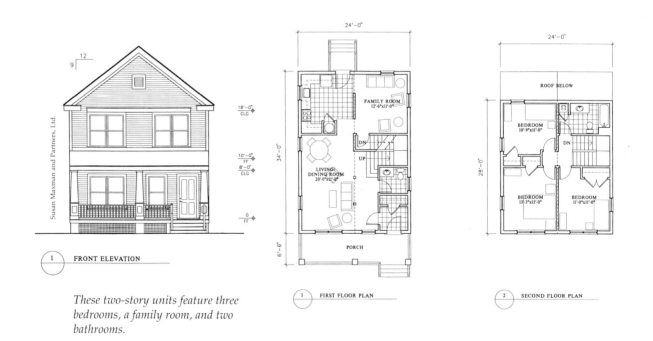

1 FRONT ELEVATION

These two-story units feature three bedrooms, a family room, and two bathrooms.

1 FIRST FLOOR PLAN

2 SECOND FLOOR PLAN

THE FUTURE

In 1996, HUD hired Steven Winter Associates, Inc., an architectural and engineering firm, and launched the Next Generation of Manufactured Housing project (NextGen) to explore future design and construction techniques. Working with an industry advisory group, Steven Winter Associates was asked to identify alternative and innovative methods, materials, and building systems that can be used in the design and construction of affordable manufactured housing nationwide.

The NextGen project developed both single-section and multisection prototypes that employ materials and construction techniques that are not typically found in manufactured housing. The single-section prototype includes add-on porches and garages, a windowless side wall to allow for private outdoor areas and zero lot line siting, a duplex unit, and the use of energy efficient, structural insulated panels (SIPs) for the exterior wall and roof portions of the homes.

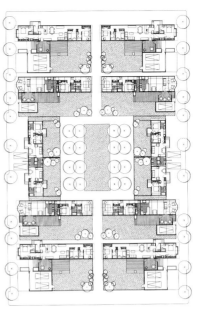

NEXT GENERATION

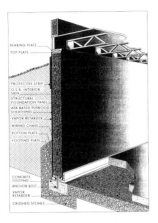

This cutaway section of the foundation of the proposed NextGen multisection home illustrates the use of structural insulated panels (SIPs). These energy efficient wall panels, consisting of a rigid foam core that is enclosed between two exterior skins, are also recommended for exterior walls and the roof in both single- and multisection homes.

Single-section homes with garages and single-section duplex homes are combined in this site plan developed by the NextGen project.

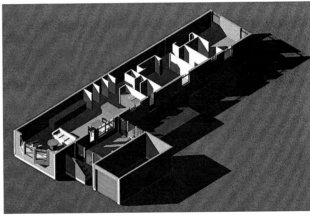

The appearance of these single-section prototypes is enhanced by add-on porches and garages. A windowless side wall allows for private outdoor areas and zero lot line siting.

Plan Options

- 3 Bedroom
- Zero-Lot-Line
- Attached Screened Porch
- Attached 2-Car Garage

The multisection prototypes include lift-up roof systems that allow for an unfinished second floor for future expansion and offset sections, or "floors" that create a porch at the front of the unit and a deck at the rear of the unit. Designed for urban infill, the multisection prototypes can also be built in a zero lot line configuration. A SIPs basement wall system has been designed for the multisection units. Among the many innovations that are also being recommended for use in both the single-section and multisection manufactured units are: lightweight interior steel studs, which save on fabrication time; glued trusses; special construction and drywall adhesives for greater durability; mechanical ventilation systems that provide continuous ventilation; and easy-to-install flexible gas piping.

This double-section prototype includes a lift-up roof system that allows for an unfinished second floor for future expansion and offset sections or "floors" that create a porch at the front of the home and a deck at the rear of the home.

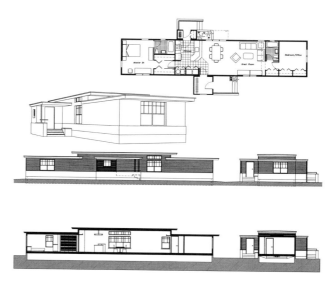

The NextGen project explored the use of a nearly flat roof. "For practical reasons," explained project personnel, "it is much easier to build a flat SIP roof than a pitched one, and a flat roof allows a high ceiling while maintaining low road clearances. . . . [T]he design also had an appealing visual character, reminiscent of Prairie School architecture and free of the often unconvincing attempts to make a single-section unit look like a normal site-built house by pitching the roof. However, the industry is struggling to overcome the highly negative public image of the typical almost-flat-roofed single-section 'mobile home'. . .any design with a nearly flat roof seems to be (and would be perceived by the public to be) a major step backward." At this point in time, the manufactured housing industy is itself very reluctant to embrace this design. Nevertheless, in some parts of the country, the Prairie School design, making use of quality materials and attention to design details may be as acceptable and as aesthetically appealing as single-section homes with pitched roofs and attached garages. The Rosa Vista development is Mesa, Arizona, designed by Andres Duany, uses units similar to these.

Goldman Sachs. 1996. *Manufactured Housing: Upwardly Mobile.* New York.

"Manufactured Housing Now." 1996. *Urban Land.* January. (entire issue)

Maxman, Susan, and Martin Muscoe. 1997. "Manufactured Housing Urban Design Project." *Urban Land.* March.

Merrill Lynch Global Securities Research and Economics Group. 1995. *The Manufactured Housing Community: The American Dream, Re-engineered.* New York.

Planning and Design Institute. 1996. *The Midtown Triangle Neighborhood Plan: Urban Design and Development of a Traditional Milwaukee Neighborhood.* Prepared for the city of Milwaukee and the Neighborhood Renewal Foundation, Milwaukee, Wisconsin.

Ryhn, Douglas. 1995. *Rethinking Manufactured Housing: A Graduate Level Design Studio Report.* Milwaukee: The School of Architecture and Urban Planning, University of Wisconsin-Milwaukee, Fall.

Sanders, Welford. 1997. "Current Trends in Factory-Built Housing." *Urban Land.* March.

_____. 1993. *Manufactured Housing Site Development Guide.* Planning Advisory Service Report No. 445. Chicago: American Planning Association. April.

_____. 1986. *Regulating Manufactured Housing.* Planning Advisory Service Report No. 398. Chicago: American Planning Association. December.

Shen, Guoqiang, and Richard A. Stephenson. 1997. *The Impact of Manufactured Housing on Adjacent Site-built Residential Properties in North Carolina.* Greenville, N.C.: Department of Planning, East Carolina University. June.

"SilverCrest Unveils New HUD-Code, Two-Story Home." 1997. *Automated Builder,* August.

Steven Winter Associates. 1997. *Next Generation of Manufactured Housing: Design Phase.* Prepared for the U.S. Department of Housing and Urban Development. Norwalk, Conn.

University of Illinois at Urban–Champaign. 1996. *Permanent Foundations Guide for Manufactured Housing.* Prepared for the U.S. Department of Housing and Urban Development, Office of Policy Development and Research. Urbana, Ill.: School of Architecture/Building Research Council. September.

Vermeer, Kimberly, and Josephine Louie. 1997. *The Future of Manufactured Housing.* Cambridge, Mass.: Joint Center for Housing Studies of Harvard University. January.

Warner, Kate, and Jeff Scheuer. 1993. *Manufactured Housing Impacts on Adjacent Property Values.* Ann Arbor, Mich.: College of Architecture and Urban Planning, University of Michigan.

Selected Bibliography

**Sample
Manufactured Housing
Provisions**

When drafting zoning and subdivision standards for manufactured homes or updating existing requirements, local officials should keep in mind that affordability is one of the most important attributes of this form of housing. Appearance standards and other requirements designed to make manufactured housing more attractive and compatible with site-built housing should be formulated with some understanding of the subsequent impact that these requirements will have on housing costs. Otherwise, the savings offered by manufactured housing can be diminished. For example, appearance standards that include minimum unit-width or floor-area requirements can restrict the use of the more affordable single-section homes, regardless of how these homes may otherwise be designed. A minimum unit-width requirement alone would restrict the use of two-story single-section homes and those single-section homes with attached garages and other additions that can enhance the appearance of these homes and make them more compatible with site-built homes. Local governments will want to continue to restrict the use of basic single-section homes, but can allow an exception when certain enhancements are provided.

A basic understanding of the HUD Code—its regulatory scope, how it compares to model building codes for site-built housing, and how code compliance is monitored and enforced—will help local officials develop reasonable and effective zoning and subdivision provisions for manufactured housing. What local officials know about how these homes are built and installed on a site may determine whether they are allowed in a community at all, and, if they are, where they will be allowed and under what conditions.

It is important that local officials know that the HUD-administered, National Manufactured Home Construction and Safety Standards preempt state and local regulations and, as noted in Part 1 of this report, HUD has taken steps in recent years to help ensure that manufactured homes are treated fairly. Strictly speaking, the language of local regulations should not be inconsistent with that of the HUD Code. It is also important that local officials keep abreast of any major changes that may occur in these standards.

Manufactured homes and manufactured home developments should be recognized as residential developments that belong in residential districts where dwellings of similar density and intensity are permitted. Restricting manufactured homes to special manufactured home districts can add to development costs by making a costly rezoning necessary prior to each new development. These costs will, of course, be passed on to the consumer. If an ample supply of land suitable for residential development is not zoned for the special manufactured home district, new development may be discouraged altogether.

Local officials who are amending or creating manufactured housing regulations should keep in mind the following suggested guidelines, which are based on the survey findings and discussions with planners and industry representatives.

The definition of manufactured housing should be clear, concise, and as consistent as possible with the HUD Code. Local ordinance definitions should clearly distinguish between different types of factory-built housing and how they will be regulated. Mobile homes, built prior to enactment of the HUD Code, should be distinguished from manufactured homes built in compliance with the code. Local provisions should also distinguish between the manufactured home and modular, panelized, and other factory-built units that must comply with state or local building codes rather than federal regulations.

Standards for manufactured home developments should be flexible so that the most cost-effective site-design opportunities are possible. Some development parameters have been suggested in Part 3 of this report, but some standards—minimum size of development, unit setbacks, perimeter and street width—may be best determined on a case-by-case basis. Flexibility recognizes that each development parcel has its own constraints and opportunities that dictate appropriate unit design and siting. Developers need to be able to vary lot dimensions, unit setbacks, and the design of the street system to best fit the site.

When specific minimum requirements are set in advance of development, design opportunities can be limited. Some of the communities featured in this report were aware of this fact and, consequently, did not set minimum or maximum requirements in areas in which they believed developers could make the best choice. If the communities did set standards they allowed some flexibility in their application.

Local regulations for manufactured home developments should include provisions that encourage good design. The importance of perimeter buffering and screening, common open space, and landscaping in manufactured home developments has been discussed. There are other site design techniques that can be used to further enhance the appearance and livability of small-lot developments like manufactured home land-lease communities and subdivisions. Some of the techniques that should be permitted and encouraged in manufactured home developments are:

1. *Elimination of one side yard and the siting of dwelling units on a side lot line.* A good way to take advantage of the limited space available on the small lots typically used in manufactured home developments is to eliminate one side yard and place the unit on a side lot line. This siting arrangement, common referred to as zero lot line (ZLL) siting, concentrates space and opens up small, narrow lots. Access to the rear of the home from the front yard and separation between homes is provided just as well by one side yard as by two. Building separation requirements can be used to ensure that the elimination of one side yard will not result in homes sited to close together. A number of the manufactured home prototypes developed by the Next Generation of Manufactured Housing project, featured in Part 4, were designed to allow for zero lot line siting.

2. *Integration of indoor and outdoor areas.* This can be achieved by designing and siting manufactured homes so that living areas within the units open out into private open space, giving the units a more spacious feeling.

3. *Variations in the exterior design and siting of homes.* A variety of floor plans and home designs with varied facade treatments should be encouraged in manufactured home development to enhance the streetscape—the visual quality of the development from the street—and to avoid a monotonous appearance. The use of identical unit designs and/or facade treatments on adjacent lots should, in most cases, be discouraged.

4. *Cluster site plans should be permitted and encouraged.* This cost-effective approach uses lots that are smaller than those typically required in a given zoning district and concentrates homes on the most buildable portion of the development site. The land saved from each individual lot is assembled to create common open space. The cluster plan keeps utility runs to a minimum, and materials and construction costs for expensive street pavement, sidewalks, curbs and gutters are reduced. One of the communities studied, Issaquah, Washington, encourages the use of cluster site plans in manufactured home subdivisions:

   > **Cluster encouraged.** A cluster development within a manufactured home subdivision shall be encouraged in order to provide affordable housing through the provision of smaller lots. A cluster development is encouraged to have fifteen percent (15%) usable open space which does not include critical areas or required buffers.

The cluster site plan is especially appropriate in areas where significant natural features or other circumstances dictate that homes should not cover the entire site. In built-up urban areas, a traditional neighborhood plan that follows the existing street pattern may be more appropriate. A neighborhood with a traditional development plan was selected in Milwaukee, Wisconsin, as the site for the city's urban design project (see pages 65-68 above).

A range of development options should be offered so that developers are able to select the regulatory approach that best suits the development proposal. This study has shown that with proper planning and regulation, and attention to design details, manufactured housing can be an attractive and affordable alternative to more costly site-built housing.

The provisions outlined below are from the zoning ordinances and development codes of the communities that responded to APA's 1996 survey. Included are both general provisions for manufactured home developments that are permitted in standard zoning districts and provisions for special districts for manufactured home developments. In most cases, however, these provisions can be used interchangeably; that is, they can be used whether a special district is established or manufactured home developments are permitted in existing zoning districts.

This appendix is divided into two parts. Part 1 includes excerpts from various ordinances. These provisions are organized into five sections that are typically found in zoning ordinances. The sections are: Title; Purpose/Intent; Development Standards; Site Plan Review/Permitting Requirements; and Definitions. Each section begins with a statement of purpose and a commentary on the provisions that follow. Part 2 includes a complete set of provisions from the Springfield, Missouri, zoning ordinance, which we believe offers an excellent example of good manufactured housing provisions.

Before deciding whether to use any of these provisions, local officials are urged to first determine if the provisions or regulatory approaches they would like to adopt are consistent with and permitted by applicable state statutes. For example, definitions should be consistent with those in state regulations.

These provisions are provided for illustration only and may be inappropriate or require significant modification to best address local circumstances.

Part 1. Excerpts from Local Ordinances

I. TITLE

Purpose. *To give a brief description of the ordinance.*

Notes. *The title should denote the scope of the provisions, indicating, for example, if the provisions apply to manufactured housing land-lease communities, subdivisions, or both. The title should also indicate whether the provisions create a special district or overlay zone. Communities can also began to update the overall language of the ordinance with the title. Many communities continue to refer to manufactured housing developments with titles like "Mobile Home Park" rather than "Manufactured Home Community" or "Development." The latter title is consistent with the language of the HUD Code.*

For General Provisions

Manufactured Home Development
   [Winston Salem–Forsyth County, North Carolina, Unified Development Codes]

Manufactured Homes and Manufactured Home Subdivisions
   [City of Issaquah, Washington, Land Use Code, 1996]

Manufactured Home Development Standards
   [Albany, Oregon, Development Code, 1994]

Manufactured Housing
   [Crown Point, Indiana, Zoning Code, 1982]

For Districts

Manufactured Home Community District
  [Springfield, Missouri, Zoning Ordinance, 1995]

Manufactured Housing District
  [Burnsville, Minnesota, Zoning Ordinance, 1990]

Manufactured Home (Overlay) Zones
  [City of Grand Island and Hall County, Nebraska, Zoning Ordinance]

Single-Family and Manufactured Home District
  [Brookings, South Dakota, Zoning Ordinance, 1995]

Planned Residential Manufactured Home Subdivision District
  [Johnson County, Kansas, Zoning and Subdivision Regulations, 1996]

II. PURPOSE AND INTENT

Purpose. *The goals and objectives of the provisions are established in this section of the ordinance.*

Notes. *The purpose statement describes how the community envisions manufactured housing development, usually in terms of overall design, density, and amenities. Including a statement of purpose in provisions for manufactured homes is especially useful to make it clear to both developers and the public the intent behind the provisions and the public interest to be served. For example, a number of communities have used the purpose statement to indicate that a key objective in permitting and encouraging manufactured home development is to allow for more affordable housing. The purpose statement can help clarify other issues as well, such as what the community considers appropriate density and the need for common open space in manufactured housing developments.*

For General Provisions

7.4.7.1 Purpose. These standards ensure that manufactured homes and subdivisions are planned, developed, and maintained to provide safety for their residents while ensuring that manufactured homes are compatible with the scale and character of the surrounding neighborhood.

7.4.7.2 Intent. The intent of these provisions is to provide affordable and diversified housing opportunities within the city while maintaining established standards. Manufactured homes shall meet the most current HUD Code standards.

[Issaquah, Washington, Land Use Code, Sections 7.4.7.1–2, 1996]

(a) Purpose. The purpose of this section is to regulate the permanent installation of manufactured homes on foundations for occupancy as single-family dwellings in accordance with and as defined in Section 64852.3 of the California Government Code and Section 18300 of the California Health and Safety Code. All such manufactured homes shall be designed and located so as to be compatible with neighboring conventionally built dwellings. The specifications provided by this section are designed to ensure the compatibility of manufactured homes in single-family zones with the aesthetic and architectural character of the surrounding neighborhood, in the same manner as that used by the county to approve other building permits for dwellings.

[Santa Cruz County, California, Code, Section 13.10.682, 1994]

Sec. 19–1. Purpose. These regulations are intended to be in accordance with the city's comprehensive plan and are designed to lessen traffic congestion; secure safety from fire; promote health and general welfare; provide adequate light and air; prevent the overcrowding of land; avoid undue concentration of population; and facilitate the adequate provision of transportation, water, sewage, schools, parks, and other public services.

[Poplar Bluff, Missouri, Code of Ordinances, Section 19.1, 1982]

(A) Intent. Manufactured homes provide a viable and affordable housing option for a segment of the county's population. This housing option is provided in areas predominately of agricultural and forest use with minimal requirements, consistent with the state code. This option is also provided under certain design criteria in more residentially developed areas where they will not conflict with developments planned for site-built dwellings.

[Roanoke County, Virginia, Use Standards, Section 30–82–5]

For Districts

Sec. 36–27. M and MD—Manufactured Home Zones

Purpose. To provide for overlay zones that will permit the placement of single- or double-wide manufactured homes within either a manufactured home [land-lease development] or manufactured home subdivision, whichever the case may be, as approved for the overlay zones. Site-built dwelling units and modular homes are also permitted within the overlay zones. A variety of densities are possible depending upon the base zone to which the overlay zones are applied.

[Grand Island and Hall County, Nebraska, Zoning Ordinance Sec. 36–27]

MP—Manufactured Home Park Residential District
Section 1. Intent and Purpose of District. It is the intent and purpose of the "M–P" Manufactured Home Park Residential District [land-lease development] to provide low-density manufactured home [land-lease development] which would be compatible with the character of the surrounding neighborhood and would be consistent with the future land-use plan of the community.

[Hays, Kansas, Zoning Ordinance, 21. Article III. Section 1]

Sec. 50.12.000 Residence RMH Single Family and Manufactured Home District
Intent. This district is intended to provide for areas of residential use with a gross density of approximately six dwelling units per acre or less. This district permits single-family dwellings, two-family dwellings, modular homes, manufactured homes, and supportive community facilities such as parks, playgrounds, schools, libraries, and churches. (Required) Acceptable similarity exterior appearance standards will insure compatibility with surrounding uses.

[Brookings, South Dakota, Zoning Ordinance, Section 50.12.010, 1995]

Article 12. Section 6. Planned Residential Manufactured Home Park District (PRMHF) and Planned Residential Manufactured Home Subdivision District (PRMHS)
A. Purpose. The zoning of property to the Planned Residential Manufactured Home Park District, (PRMHP) [land-lease development] district is intended to accommodate the grouping of manufactured home sites for use on lots rented or leased to the occupant of the manufactured home. The zoning of property to the Planned Residential Manufactured Home Subdivision District (PRHMS) is intended to accommodate manufactured homes on subdivided lots deeded to individual property owners. These districts are intended to provide a safe and healthy living environment and to ensure the mutual compatibility of Manufactured Home [land-lease development] and Manufactured Home Subdivisions with adjoining land uses.

[Johnson County, Kansas, Zoning and Subdivision Regulations, 1996]

14–6D–6: Factory-Built Housing Residential Zone (RFBH)
Intent. The Factory-Built Housing Residential Zone (RFBH) is designed to provide for the placement of manufactured homes, mobile homes and modular homes in factory-built housing [land-lease development] or on individually subdivided lots with a lot size smaller than that allowed in other zones permitting single-family dwellings.

The RFBH zone also provides a location for the placement of those mobile homes which are not classified as manufactured homes and may not comply with the Building, Electrical, Plumbing or Housing Codes all as amended, and for those factory-built homes which do not have a minimum building width of 20 feet. In addition, the zone provides for integration of limited commercial and service uses to serve the residents of the zone.

[Iowa City, Iowa, Zoning Ordinance, Section 14.6D–6. 1994]

Sec. 138–1343. Residential Design Manufactured Homes

(a) Intent. It is the intent of this section to encourage the provision of affordable housing in a general residential environment by permitting the use of residential design manufactured housing (RDMH), as defined in this section, in residential districts in which similar dwellings constructed on the site are permitted, subject to the requirements and procedures set forth in this section to assure similarity in exterior appearance between such residential designed manufactured housing and dwellings which have been constructed under these and other lawful regulations on adjacent lots in the same district. Manufactured homes approved as RDMH, either individually or by specific model, shall be permitted in residential districts in which similar residential occupancy is permitted, subject to requirements and limitations applying generally to such residential use in the districts, including minimum lots, yards, and building spacing, percentage of lot coverage, off-street parking requirements, and approved foundations, as described in this chapter.

[Pinellas County, Florida, Zoning Ordinance, 1992]

III. DEVELOPMENT STANDARDS

Purpose. *To establish minimum or maximum development parameters for manufactured home developments.*

Notes. *This section includes two categories of development standards that are found in zoning ordinances—standards that apply when manufactured homes are placed on individual lots and provisions for manufactured home developments, including land-lease communities and manufactured home subdivisions. Development standards for manufactured homes that may be placed on individual lots in existing residential or built-up residential areas usually include appearance standards to help ensure that these homes will be compatible with site-built homes.*

In some cases, a single set of requirements has been established for both land-lease communities and subdivision developments. More common, however, is to establish a set of standards for manufactured home land-lease communities only and require that manufactured home subdivisions adhere to the same requirements as subdivisions involving site-built homes. A third variation is to have different standards for each.

Typically, development standards for manufactured housing land-lease communities and subdivisions have been established in zoning ordinances in at least three different ways—as an overlay district, as a special zoning district, or through the application of general provisions for manufactured home developments in existing residential districts.

On Individual Lots

10.170 Manufactured Home Placements. Manufactured homes are permitted on individual parcels or lots outside of manufactured home [land-lease development] in accordance with the placement standards set forth in Sections 10.100 and 10.120 and all other provisions of the Development Code for site-built dwellings.

10.180 Review Criteria. In order to be approved, the manufactured home must be found to have design compatibility with other dwellings in the "review area," which is the area within 300 feet of the subject lot or parcel or the nearest five dwellings. The criteria for determining acceptable compatibility shall be based upon a review of the following design elements:

(1) Roofing shall be similar in color, material, and appearance to the roofing material commonly used on residential dwellings within the community or comparable

to the predominant materials used on dwellings within the review area. The roof pitch shall be a minimum of nominal 3/12. Manufactured homes placed in RM–5 or RM–3 districts may have a roof pitch of nominal 2/12.

(2) Exterior siding shall be similar in color, material, and appearance to the exterior siding material commonly used on residential dwellings within the community or comparable to the predominant materials used on dwellings within the review area.

(3) A garage of like materials and color of the attached dwelling is required where similar features are predominant in the review area. A carport may be allowed if other dwellings in the review area also have carports or if there is a mixture of dwellings with or without garages or carports. The garage or carport may be required to be attached if other dwellings in the review area have attached garages.

(4) All Class A and Class B manufactured homes outside of manufactured home parks shall be placed on an excavated and back-filled foundation and enclosed with a perimeter enclosure, which must be similar in appearance to foundations or enclosures in the area.

[Albany, Oregon, Development Code, 1994]

119.710. Manufactured Homes on Individual Lots. Where permitted as a special use, manufactured homes on individual lots shall meet the following non-variable development standards:

(a) The manufactured home shall be multisectional and enclose a space of not less than 860 square feet.

(b) The manufactured home shall be placed on a excavated and back-filled foundation, and enclosed continuously at the perimeter with material comparable to the predominant materials used in foundations of surrounding dwellings.

(c) The manufactured home shall have a pitched roof, with a slope not less than a nominal three feet in height for each 12 feet in width.

(d) The manufactured home shall have exterior siding and roofing which in color, material, and appearance is similar to the exterior siding and roofing material commonly used on residential dwellings within the community or which is comparable to the predominant materials used on surrounding dwellings as determined by the city.

(e) The manufactured home shall be certified by the manufacturer to have an exterior thermal envelope meeting performance standards equivalent to the performance standards required of single-family dwellings constructed under the state building code as defined in ORS 455.010.

(f) The manufactured home shall have a garage or carport constructed of like materials.

The nonvariable nature of the foregoing development standards means that said standards may not be varied or deleted by the conditional use approval process.

[Salem, Oregon, Zoning Code 1995]

Section 9.08.110 Manufactured Home Requirements
A. General Provisions. Individual manufactured homes may be permitted on individual lots in the HR, RR, RI, RA2, R2, R3 and R5 Districts, subject to the following requirements.

1. The structure is placed on a permanent foundation in compliance with all applicable building regulations.

2. The structure is certified under the National Manufactured Housing Construction and Safety Standards Act of 1974 and was constructed not more than 10 years prior to request to install. Documentation indicating certification and construction date must be submitted to the Community Development Department in order to secure valid building permit(s).

3. The Community Development Director shall determine if placement of the manufactured home is compatible with the immediate area and meets the development standards of the underlying district. Installations shall be subject to the minor development review process (Section 9.02.030 A) and the following design criteria:

   a. The design of the structure shall be similar in character and appearance to other dwellings in the area with regard to unit size, roof overhangs, roof materials, roof pitch, and exterior materials;

   b. All building setbacks, parking, coverage, height, width, and sign requirements of the base district shall apply;

   c. A roof constructed of asphalt composition, shingle, tile, crushed rock, or similar roofing material (except metal) which is compatible with surrounding development;

   d. Exterior siding of brick, wood, stucco, plaster, concrete, or other material which is finished in a nonglossy and nonreflective manner and which is compatible with surrounding development;

   e. A predominant shape and form that is compatible with the surrounding neighborhood; and

   f. If an enclosed garage is required within the zoning district in which the dwelling unit is to be located, the design and materials of the garage shall be compatible with the main dwelling.

[Moreno Valley, California, Development Code]

Section 28–48. Architecture

(a) Generally. In all districts, all principal buildings other than one- and two-family districts must be designed or approved by a registered architect or engineer. A building permit shall not be issued where the materials, scale, bulk, or character of a structure, house, or building is so similar or dissimilar to other structures, houses, or buildings in the vicinity as to result in depreciation of property values or the degradation of the environment in the area. However, any denial of such a permit must be confirmed by the city council.

(b) Residential districts. In districts R–1, R–2, R–3, R–4, R–5 and R–6, the following standards shall apply for single-family detached dwellings:

   (1) **Minimum width.** The minimum width of the main portion of the structure shall be not less than 20 feet, as measured across the narrowest portion.

   (2) **Minimum roof pitch.** The pitch of the main roof shall be not less than 2.5 feet of rise for each 12 feet of horizontal run. This requirement may be waived for earth-sheltered structures.

   (3) **Placement.** Every single-family dwelling shall be placed so that the apparent entrance or front of the home faces or parallels the principal street frontage, except where the lot size exceeds one acre.

   (4) **Foundation.** Every dwelling shall be placed on a permanent foundation in compliance with the Uniform Building Code as adopted by the city.

[Cottage Grove, Minnesota, City Code 1994]

155.04 Exterior Appearance Standards. A manufactured home shall comply with the following:

(A) It shall conform to the minimum square footage requirements of appropriate zoning classification in the zoning code for residential construction.

(B) It shall be placed on a foundation meeting all the requirements of a conventional single-family home foundation.

(C) It shall be anchored to the ground, in accordance with approved manufactured home installation standards.

(D) It shall have wheels, axles, and hitch mechanisms removed.

(E) It shall meet appropriate utility connection standards, in accordance with approved home installation standards.

(F) It shall have siding material of a type customarily used on site-constructed residences.

(G) It shall have roofing materials of a type customarily used on site-constructed residences, with a minimum of three inches of pitch for each foot of horizontal travel.

(H) It shall be placed onto a permanent exterior perimeter retaining wall, in accordance with the approved manufactured home installation standards.

155.05 Installation Standards

(A) **Perimeter retaining wall.** Those manufactured homes designated in the zoning code as requiring perimeter retaining walls must be set onto an excavated area with foundations, footing, and basement walls constructed in accordance with the terms of the one- and two-family dwelling code. The space between the floor joists of the home and the excavated under floor grade shall be completely enclosed with the permanent perimeter retaining wall. The wall shall be composed of solid masonry, which shall extend below the frost line. The design by a registered professional engineer or architect shall safely support those loads, as determined by the character of the soil.

(B) **Foundations.** All manufactured housing shall be installed on foundations in accordance with the requirements of the one- and two-family dwelling code of the state.

[Crown Point, Indiana, Zoning Code, 1982]

17.90.065 Architectural Design Elements. All single-family dwellings, modular housing, and manufactured homes located in the RVL, RL, RI, RZ, RAM, R3, RM, and C1 zones shall utilize at least two of the following architectural features: dormers; more than two gables; recessed entries; covered porch/entry; bay window; building off-set; deck with railing or planters and benches; or a garage, carport, or other accessory structures.

[Cannon Beach, Oregon, Zoning Ordinance, 1995]

In Manufactured Home Developments

1. Overlay Zones

30–4–4.5 Manufactured Housing Overlay Districts

(A) **Authority.** Manufactured Housing Districts may be established as authorized by NCGS Article 19, Section 160A.383.1(e) and this Ordinance.

(B) **Minimum Criteria.** Class AA manufactured dwellings may be permitted on single-family lots in any residential district provided overlay district zoning is approved by the city council and each manufactured dwelling:

1) is occupied only as a single-family dwelling;

2) has a minimum width of 16 feet;

3) has a length not exceeding four times its width, with length measured along the longest axis and width measured perpendicular to the longest axis at the narrowest part;

4) has a minimum of 1,000 square feet of enclosed and heated space;

5) has the towing apparatus, wheels, axles, and transporting lights removed and not included in length and width measurements;

6) has the longest axis oriented parallel or within a 10 degree deflection of being parallel to the lot frontage, unless other orientation is permitted as a special exception by the Board of Adjustment following a public hearing;

7) is set up in accordance with the standards established by the NC Department of Insurance. In addition, a continuous, permanent masonry foundation or

masonry curtain wall constructed in accordance with the standards of the NC Building Code for One- and Two-Family Dwellings, unpierced except for required ventilation and access, shall be installed under the perimeter;

8) has exterior siding comparable in composition, appearance, and durability to the exterior siding commonly used in standard residential construction, consisting of one or more of the following:

   a) Vinyl or aluminum lap siding (whose reflectivity does not exceed that of flat white paint);

   b) Cedar or other wood siding;

   c) Wood grain, weather-resistant, press board siding;

   d) Stucco siding; or

   e) Brick or stone siding

9) has a roof pitch minimum vertical rise of 3.5 feet for each 12 feet of horizontal run;

10) has a roof finished with a Class C or better roofing material that is commonly used in standard residential construction;

11) has an eave projection for all roof structures of no less six inches, which may include a gutter; and

12) has stairs, porches, entrance platforms, ramps, and other means of entrance and exit installed or constructed in accordance with the standards set forth in the NC Building Code, attached firmly to the primary structure and anchored securely to the ground. Wood stairs shall be used in conjunction only with a porch or entrance platform with a minimum of 24 square feet.

(C) **Minimum Development Size:**

(1) Ten existing contiguous lots covering at least 90,000 square feet, excluding public street right-of-way; or

(2) One hundred and twenty thousand square feet of land, excluding public street right-of-way.

(D) **Additional Information:** In requesting the establishment of an overlay district, an applicant shall present factual information to ensure, in the opinion of the City Council, that property values of surrounding properties are protected, that the character and integrity of the neighborhood are adequately safeguarded, and the purposes of the overlay district as set forth in Section 30–1–3.7 (Manufactured Housing overlay Purposes) are met. To ensure acceptable similarity in exterior appearance between proposed manufactured dwellings and dwellings that have been constructed on adjacent or nearby lots, an applicant may, for illustrative purposes, present examples of the types and design of such proposed dwellings.

[Greensboro, North Carolina,Development Ordinance, Section 30–4–4.5]

Manufactured Home Overlay

Section 10.401. Purpose. The purpose of the Manufactured Home Overlay is to provide for the development of manufactured housing in established residential zoning districts while maintaining the overall character of those districts. The intent of the prescribed conditions herein is to ensure compatibility with existing housing stock through aesthetically related standards. The Manufactured Home Overlay shall be an overlay in any districts permitting residential development (R–3, R–4, R–5, R–6, R–8, R–8MF , R–12MF, R–17MF, R–22MF, R–43MF, 0–1, 0–2, 0–3, B–1, B–2, MX–1, MX–2 and MX–3) established in Chapters 9 and 11, except UR and UMUD districts. The Manufactured Home Overlay supplements the range of uses permitted in the underlying district. All other uses and regulations for the underlying district shall continue to remain in effect for properties classified under the Manufactured Home Overlay.

Section 10.402. Procedures for District Designation;
Additional Application Content Requirements

(1) A Manufactured Home Overlay district shall only be designated for a contiguous area of at least two acres in size.

(2) Property shall be classified under the Manufactured Home Overlay district only upon a petition filed by an owner of the property, or anyone else authorized in writing to act on the owner's behalf, and approved by the City Council under the procedures and standards established in Chapter 6, Part 1, of these regulations. Uses in the Manufactured Home Overlay district shall be subject only to those conditions and standards established in this Part and those conditions and standards established in Chapters 9 and 11, for uses permitted in the underlying district. All uses permitted in the underlying district are permitted in the Manufactured Home Overlay district.

(3) The petition shall be accompanied by the following information: A site plan is required. The site plan shall acknowledge and demonstrate compliance with the prescribed conditions described in Section 10.403. Once an overlay district is approved, a building permit shall not be issued until the site plan has been approved by the Planning Director.

(4) Following the City Council designation and approval of a Manufactured Home Overlay district, the area so designated shall be labeled 'MH–0" on the Official Zoning Maps.

Section 10.403. Uses Permitted Under Prescribed Conditions. The following uses shall be permitted as of right in the Manufactured Home Overlay district provided that they meet the standards established in this Section and all other requirements of these regulations.

(4) Manufactured homes, in accordance with the following standards:

(a) The home shall be set up in accordance with the standards set by the North Carolina Department of Insurance, and a continuous, permanent, masonry wall, having the appearance of a conventional load-bearing foundation wall, unpierced except for required ventilation and access, shall be installed under the perimeter of the manufactured home;

(b) The home will have all wheels, axles, transporting lights, and towing apparatuses removed;

(c) The structure must be at least 24 feet in width along the majority of its length. However, within an underlying R–8 district, the width may be reduced to 22 feet;

(d) All roof structures will have a minimum nominal 3/12 pitch and must provide an eave projection of no less than six inches, which may include a gutter. The roof must be finished with a type of shingle commonly used in site-built residential construction;

(e) Exterior wall materials and finishes must be comparable in composition, appearance, and durability to those commonly used in standard residential construction. Vinyl and aluminum lap siding, wood, stucco, brick, and similar masonry materials may be used. Reflectivity shall not exceed that of gloss white paint; and

(f) All entrances to a manufactured home shall be provided with permanent steps, porch, or similar suitable entry.

(2) All principal and accessory uses in the underlying district are permitted.

[*Editor's Note: Manufactured homes that do not comply with the standards established for homes permitted in overlay districts may be permitted in R–MH Districts, subject to the following requirements.*]

Section 11.304. Development Standards; Density; Common Area Requirements.
All uses and structures in the R–MH District shall meet the development standards established in Section 11.307 of this Part, and the following:

(1)   A manufactured home [land-lease development] or mobile home park or subdivision located within the R–ME district shall be at least two acres in area and the maximum size allowed for any rezoning to the R–MH district is 40 acres.

(2)   No structure shall be located within 30 feet of any property line defining the perimeter of the manufactured home [land-lease development] or mobile home park or except as otherwise provided in subsection 12.106.

(3)   Each lot or space within the [land-lease development] shall be at least 5,000 square feet in area and at least 40 feet wide. No more than one home may be erected on one space. In a subdivision, the lot and yards shall be developed to the standards of the R5 district.

(4)   Any structure shall be located at least 20 feet from any internal street and at least 10 feet from any adjacent lot or space within the [land-lease development] or subdivision except as otherwise provided in Section 12.106.

(5)   The overall density of homes within the [land-lease development] or subdivision shall not exceed six units per acre.

(6)   There must be at least four spaces available at first occupancy in a manufactured home [land-lease development] or mobile home park.

(7)   All manufactured and mobile homes, buildings, and service areas will be separated by a Class C buffer from any abutting property located in a residential district or abutting residential use.

(8)   At least 8 percent of the total area of a manufactured housing [land-lease development] shall be devoted to recreational use by the residents of the [land-lease development]. Such use may include space for community buildings, gardens, outdoor play areas, swimming pools, ball courts, racquet courts, etc.

(9)   No service building, office, or common recreational area shall be located adjacent to a public street or any property line defining the perimeter of the [land-lease development] or subdivision.

Section 11.305. Streets and Utilities
(1)   Each lot or space shall be equipped with electricity, drinking water, and wastewater disposal facilities.

(2)   A park shall be equipped with paved private streets built to the specifications of the "Charlotte–Mecklenburg Land Development Standards Manual."

(3)   A subdivision shall be equipped with paved public streets built to the specification of the 'Charlotte–Mecklenburg Land Development Standards Manual."

(4)   Internal streets and circulation patterns shall be adequate to handle the traffic to be generated by the development.

Section 11.306. Foundations, Patios, and Walkways
(1)   Each home shall be placed on a permanent stand in accordance with standards set by the North Carolina Department of Insurance.

(2)   Each home shall have an area on site for provision of a permanent patio or deck adjacent or attached to the permanent stand of at least 180 square feet.

(3)   A walkway shall be constructed for each lot or space to connect parking spaces to the manufactured home entrance.

(4)   Attached structures, such as an awning, cabana, storage building, carport, windbreak, or porch, which has a floor area larger than 25 square feet and is roofed shall be considered part of the stand for purposes of all setback and yard requirements.

(5)   The area beneath a home must be fully enclosed with durable skirting within 60 days of placement in the park or subdivision. As a minimum, such skirting must

be a product designed and sold for use as skirting or as approved by the Zoning Administrator.

[Charlotte and Mecklenburg County, North Carolina, Development Code]

2. Manufactured Home Districts

2–5.43. Manufactured Housing Development

(A) **Site Size and Dimensional Requirements:**

    (1) **Minimum size.** The minimum size of a zoning lot to be used as a manufactured housing development shall be four acres for initial development.

    (2) **Minimum width.** The minimum width of a zoning lot to be used as a new manufactured housing development shall be 250feet. The site width shall be measured at the manufactured home space closest to the front lot line of the development.

    (3) **Lot size.** Each manufactured home space shall have a minimum area of 4,000 square feet with a minimum width of 40 feet for singlewide homes and a minimum area of 5,000 square feet with a minimum width of 50 feet for multisectional units.

    (4) **Setbacks.** Each manufactured home space shall meet the following setback requirements.

        (a) **Front yard.** The minimum front yard shall be 20 feet.

        (b) **Rear yard.** The minimum rear yard shall be 10 feet.

        (c) **Side yard.** Minimum side yard shall be five feet, with a combined width of both side yards of 15 feet.

(B) **Minimum number of spaces:** A manufactured home development shall contain no fewer than 10 manufactured home spaces for initial development.

(C) **Density:** The maximum density of a manufactured housing development shall not exceed five spaces per gross acre; with the exception that the maximum density may be increased to 5.5 manufactured home spaces per gross acre when at least 12 percent of the gross site area is in common recreation area.

(D) **Utilities:**

    (1) **Location.** All utilities within a manufactured home development shall be located underground.

    (2) **Water.** Connection to a public water system and installation of fire hydrants meeting the standards of the appropriate jurisdiction are required.

    (3) **Sewer.** Connection to a public sewer system or installation of an approved package treatment plant is required.

(E) **Bufferyards:** A type II bufferyard of a minimum width of 30 feet shall be established along each exterior property line, except where adjacent to a private street or public right-of-way not internal to the development. Along external private streets or public rights-of-way, a type II bufferyard of a minimum of 50 feet shall be established.

(F) **Access:**

    (1) **External access.** No manufactured home space shall have direct vehicular access to a public or private street outside the development.

    (2) **Internal access.** Each manufactured home space shall have direct vehicular access to an internal private access easement and street.

(G) **Common recreation area.** A minimum of 4,000 square feet or 100 square feet per manufactured home, whichever is greater, of common recreation area shall be provided in accordance with the standards of Section 3–6.

(H) **Manufactured Home Spaces:**

    (1) **Construction.** Each manufactured home space shall be constructed in compliance with the North Carolina Manufactured Home Code.

(2) **Patio or deck area.** A patio or deck area, constructed of concrete, brick, flagstone, wood, or other hard surface material and being a minimum of 144 square feet in area, shall be constructed within each space.

(3) **Walkway.** A hard surface walkway, being a minimum of two feet wide, leading from the major entrance of the manufactured home to its parking spaces or to the street shall be constructed.

(4) **Solid waste.** Each space shall have a minimum of one solid waste container with a tight fitting cover and a capacity of not less than 32 gallons, or dumpsters of adequate capacity may be substituted. If dumpsters are provided, each such container shall be located on a concrete slab and screened on three sides by an opaque fence at least eight feet in height.

(I) **Manufactured Homes**

(1) **Class D.** Class D manufactured homes shall not be permitted in new manufactured housing developments or expansions of existing manufactured housing developments. Existing Class D manufactured homes located in a manufactured housing development in operation at the time of adoption of this ordinance are allowed to remain, but, if removed, shall be replaced with a Class A, B, or C manufactured home.

(2) **Setup.** Each manufactured home shall meet the setup requirements of the North Carolina Manufactured Home Code.

(3) **Skirting.** Each manufactured home shall have skirting installed in accordance with the following requirements:

(a) Skirting shall be of noncombustible material or material that will not support combustion. Skirting material shall be durable and suitable for exterior exposures;

(b) Any wood framing used to support the skirting shall be of approved moisture-resistant, treated wood;

(c) The skirting shall be vented in accordance with state requirements;

(d) Skirting manufactured specifically for this purpose shall be installed in accordance with the manufacturer's specifications;

(e) Skirting shall be installed no later than 60 days after the set up of the home; and

(f) Skirting shall be properly maintained.

(4) **Additions.** Prefabricated structures built by a manufacturer of manufactured home extensions meeting United States Department of Housing and Urban Development standards and any other additions meeting the State Residential Building Code may be added to any manufactured home provided that setbacks within the space can be met and a building permit is obtained.

(5) **Vacant manufactured homes.** No storage of unoccupied and/or damaged manufactured homes is permitted.

(J) **Accessory structures and uses.** Accessory structures and uses permitted in manufactured housing developments shall meet standards in Sections 2–6 and 3–1.2(F) and (G).

(K) (*Editor's note: Omitted here.*)

(L) **Existing Manufactured Housing Developments:**

(1) Schedule for Improvements. Manufactured housing developments lawfully existing at the time of the adoption of this ordinance shall be required to meet the following standards of this section within four years of the ordinance's adoption date:

(a) **Bufferyards.** Section 2–5.43(E), with the exception of meeting minimum width requirements in developments where meeting the width requirements would result in the relocation of structures or manufactured homes.

   (b) **Solid waste.** Section 2–5.43(H)(4).

   (c) **Skirting.** Section 2–5.43(l)(3).

   (d) **Utilities.** Section 2–5.43(D), unless public water and sewer is located more than 200 feet from the manufactured home development.

   (e) **Streets.** Streets shall have a minimum of four inches of gravel and be well maintained.

(2) **Expansion of nonconforming manufactured housing developments.** No expansion of a nonconforming manufactured housing development shall be permitted unless all units in the development, both pre-existing and additional, have vertical skirting or a similar structural enclosure around the entire base of the unit between the outer walls and the ground or paved surface, and are anchored to the ground in accordance with the regulations set forth by the State of North Carolina for manufactured and modular housing units.

[Winston–Salem/Forsyth County, North Carolina
Unified Development Ordinances]

3. Existing Zoning Districts

7.4.7. Manufactured Homes and Manufactured Home Subdivisions

a. **Manufactured Home Subdivision:**

(1) **Development standards.** A manufactured home subdivision shall be subject to the same land development and site improvement standards that apply to conventional subdivisions.

(2) **Cluster encouraged.** A cluster development within a manufactured home subdivision shall be encouraged in order to provide affordable housing through the provision of smaller lots. A cluster development is encouraged to have 15 percent usable open space which does not include critical areas or their required buffers. This usable open space should have the ability to provide for recreation.

b. **Manufactured Home:**

(1) **External material.** The home shall be covered with an exterior material customarily used in conventional dwellings including, but not limited to, wood clapboards, simulated clapboards such as conventional vinyl or aluminum siding, masonry, wood singles, shakes, or similar material, but excluding smooth, ribbed, or corrugated metal or plastic panels. The exterior covering shall extend to the ground, except that when a solid concrete or masonry perimeter foundation is used, the exterior covering material need not extend below the top of the foundation.

(2) **Fuel oil supply systems.** All fuel oil supply systems shall be constructed and installed within the foundation wall or underground and within all applicable building and safety codes except that any bottled gas tanks may be fenced so as not to be clearly visible from the street or abutting property.

(3) **Mobility.** The hitch, axles and wheels shall be removed.

(4) **Permanent foundation.** The manufactured home shall be placed on a permanent foundation that complies with the city's building code for residential structures. The foundation shall be installed using trench style construction with the foundation no higher than six inches above existing grade.

(5) **Roof Pitch.** Roof pitch shall comply with the TJBC. The roof overhang shall not be less than one foot, measured from the vertical side of the manufactured home. When carports, garages, porches, or other similar structures are attached as an integral part of the manufactured home, the Planning Director/Manager may waive the overhang requirement.

(6) **Roofing materials.** Roofing materials shall consist of shingles or other material customarily used for conventional dwellings including, but not limited to, approved wood, asphalt composition shingles or fiberglass, but excluding, corrugated fiberglass or metal corrugated roof.

[Issaquah, Washington, Land Use Code, 1996]

Manufactured Homes Land Development Standards
Section 19–2. Site Requirements

(a) **Zoning.** Manufactured housing subdivisions shall be located in RS–4 or RS–5 zoning districts. Manufactured home [land-lease developments] shall be permitted as conditional uses in RS–5 zoning districts.

(b) **Permitted uses in RS–4 and RS–5 districts.** Permitted uses shall be any permissive use of the RS–2 and RS–3 districts, modular homes, and manufactured home subdivisions.

(c) Conditional uses. Conditional uses shall be any conditional use of the RS–2 and RS–3 districts, manufactured home [land-lease development].

(d) **Environment.** Conditions of soil, groundwater level, drainage, and topography shall not create hazards to the property or the health or safety of the residents. The site shall not be exposed to objectionable smoke, dust, noise, odor, or other adverse influences, and no portion subject to predictable sudden flooding or erosion and shall not be used for any purpose which would expose persons or property to hazards.

(e) **Minimum area for developments.** Subdivisions and [land-lease development] shall not contain less than five acres. A minimum depth and width for a park or subdivision shall be no less than 200 feet.

(f) **Buffering/screening.** Buffering and/or screening shall be required along the exterior boundaries of manufactured home developments that adjoin residential areas. Screening or buffering may be required in other locations when a nuisance or obnoxious use would interfere with the enjoyment of the proposed development. For example, screening may be required when a development is located near a junkyard, manufacturing plant, commercial area, or other use which is not compatible with the development.

(g) **Open space**
(1) Subdivisions in RS–4 districts shall provide no less than 5 percent of the net developable land area for common open space.
(2) Subdivisions and [land-lease development] in RS–5 districts shall provide no less than 10 percent of the net developable land area for common open space.
(3) The minimum size for a single parcel of ground for common open space for a single subdivision or park development shall not be less than 7,500 square feet.
(4) Payments in lieu of land dedication may be accepted by the city and used to develop further recreational facilities.

(h) **Sign limitations.** No sign intended to be read from any public way shall be permitted except:
(1) No more than one identification sign, not exceeding 80 square feet per side for each principal entrance to the development.
(2) No more than one sign, not exceeding six square feet per side in area, advertising property for sale, lease, or rent, or indicating "Vacancy" or "No Vacancy" may be erected at each principal entrance.
(3) All external illuminated signs shall have lights directed only toward the sign and shall be shielded by a reflector so that the light concentrates on the sign only.

(i) **Access**
(1) Access to these developments shall not be from a minor residential street.

(2) The number and location of access drives shall be controlled for traffic safety and the protection of surrounding properties.

(3) No individual lot shall be designated for direct access to a street outside the boundaries of the development.

(4) Access entrances shall be located no closer together than 400 feet (measured from center line to center line of the streets.)

(j) **Circulation**

(1) **Streets generally, parking.** Streets shall have direct connections to a public street. No parking shall be permitted for a distance of 30 feet from the corners of the beginning of the right-of-way.

(2) **Type of streets.** The development shall provide publicly dedicated streets. Required improvements shall be provided in compliance with the street paving standards described in Chapter 32 of the Code of Ordinances. Looped and cul-de-sac streets may be provided with 40-foot right-of-way and 20-foot paving widths when these streets serve less than 20 dwelling units and when off-street guest parking is provided.

(3) **Street layout.** Streets, drives, and parking areas shall provide safe and convenient access to dwellings and emergency vehicles but streets shall not be laid out to encourage outside traffic to transverse the development, nor create unnecessary fragmentation of the development into small blocks. In general, block size shall be consistent with use, the shape of the site, and the convenience and safety of the occupants.

(4) **Vehicular access to streets**. Vehicular access to streets from off-street parking may be directed from dwellings if the street serves 50 units or less. Along streets serving more than 50 dwellings or major routes to or around central facilities, access from parking and service areas shall be so combined, limited, located, and controlled as to channel traffic in a manner that minimizes traffic direction and direct vehicular access from individual dwellings shall generally be prohibited.

Section 19–3. Lot Requirements in RS4 Districts. Lot requirements in RS–4 districts shall be as follows:

(a) Minimum lot area in square feet: 6,000

(b) Minimum lot width in feet at the building line: 50

(c) Minimum street frontage in feet: 25

(d) Minimum setback requirements:

(1) Front yard in feet: 15

(2) Side yard in feet: 6

(3) Rear yard:

(a) Units placed at an angle or perpendicular to the fronting street in feet: 10

(b) Units placed parallel to the fronting street in feet: 20

Section 19–4. Lot Requirements in RS–5 Districts. Lot requirements in RS–5 districts shall be as follows:

(a) Minimum lot area in square feet: 5,000

(b) Minimum lot width in feet at building line: 45

(c) Minimum street frontage in feet: 25

(d) Minimum setback requirements:

(1) Front yard in feet: 15

(2) Side yard in feet: 6

(3) Rear yard:

a. Units placed at an angle or perpendicular to the fronting street in feet: 10

b. Units placed parallel to the fronting street in feet: 20

e. Minimum distance between buildings in manufactured home parks is 20 feet. Front yard setbacks in parks shall be no less than 15 feet.

Section 19–5. Off-Street Parking. No less than two parking spaces shall be provided to serve each dwelling unit.

Section 19–6. Lot Coverage. The maximum lot coverage by roofed or enclosed structures shall be no greater than 50 percent in [land-lease development] and subdivisions.

Section 19–7. Density

(a) The number of dwelling units per net developable land shall not exceed the following densities in the following zoning districts: RS–4, 7.2 dwelling units per acre; RS–5, 8.7 dwelling units per acre; and manufactured home [land-lease development], nine dwelling units per acre.

(b) Net developable land is defined as the gross area less public right-of-way, publicly dedicated park land, drainage ways, and unbuildable area, such as steep slopes, swamps, landfills, and unstable soil.

Section 19–8. Accessory Uses

(a) Each manufactured home lot and stand shall provide enclosed storage space in a building or in the crawl space below the living area of the manufactured home. Storage space beneath the main floor of the manufactured home will be counted as storage space if the minimum clearance between the ground (or foundation floor) and the bottom of the manufactured home is 42 inches. The area of the below building storage space shall not be less than 100 square feet. The area of the storage building shall not exceed 150 square feet and shall be no less than 24 square feet in area.

(b) Subdivisions and [land-lease development] may incorporate the following accessory uses for the benefit of development subject to the approval of the city:
  (1) Sales office/manufactured home display area
  (2) Maintenance/storage buildings
  (3) Laundry facilities
  (4) Cafeteria/restaurant
  (5) Swimming pool
  (6) Barbershop/beauty shops
  (7) Greenhouse
  (8) Game room
  (9) Sports courts
  (10) Gazebo
  (11) Daycare facilities.

[Poplar Bluff, Missouri, Code of Ordinances, 1982]

Section 33442. Requirements for Manufactured Home Subdivisions. Manufactured home subdivisions must meet the following requirements:

A. Manufactured home subdivisions are only permitted in the General District; see the Table of Permitted Principal Uses in Section 334–21 of Article V.

B. The minimum size of the tract of land to be subdivided must be 10 acres.

C. Only single-family manufactured homes are permitted, and only one manufactured home may be placed on each residential lot within a subdivision.

D. Each manufactured home shall be affixed to a permanent foundation.

E. The subdivision shall be screened along its perimeter by a permanent buffer area, not less than 50 feet wide, composed of trees, shrubs, or other suitable buffers approved by the planning board. The buffer area may be placed on individual lots within the subdivision.

F. Each manufactured home lot within the subdivision shall meet the dimensional requirement for single-family use in the Table of Minimum Dimensional Requirements in Section 334–27 of Article VII.

[Town of Hudson, New Hampshire, Zoning Ordinance, 1995]

17.43.020 Manufactured Home [Land-Lease Development]. Manufactured home [land-lease development], in the zones where permitted, shall meet the following requirements, in addition to any conditions which may be imposed by the Use Permit:

A. **Minimum area.** The minimum [land-lease development] area shall be five acres.

B. **Minimum number of sites.** The minimum number of manufactured home sites shall be 50.

C. **Density.** The density of a manufactured home [land-lease development] shall not exceed the density range as defined in the General Plan for the property on which the [land-lease development] is located.

D. **Setbacks between manufactured homes.** There shall be a distance of at least 15 feet between any manufactured home and any other manufactured home within a Manufactured Home Park.

E. **Landscaping.** A Manufactured Home [land-lease development] shall meet the landscaping requirements of the zone in which it is located as provided by the Landscaping Development Guidelines.

F. **Fencing**. A Manufactured Home [land-lease development] shall be enclosed by a masonry wall of at least seven feet in height located on the property side of the street landscape setback as defined by the Landscape Development Guidelines and along all property lines adjoining another private property.

G. **Roads.** All circulation roads within a manufactured home [land-lease development] shall be paved at least 25 feet wide from curb to curb. Ten feet additional width shall be provided if parking is to be permitted on one side of the roads, and 20 feet additional width shall be provided if parking is to be provided on both sides of the roads.

H. **Automobile parking.** There shall be an equivalent of two parking spares per manufactured home site. The remaining required-automobile parking areas shall be conveniently located in relation to office, recreation, and service areas.

I. **Paving.** All areas used for access, parking, or circulation shall be permanently paved.

J. **Access.** Each manufactured home [land-lease development] shall be so designed that access to public roads is provided to the satisfaction of the Public Works Department and Fire Department.

K. **Improvement of existing manufactured home [land-lease development]**. Upon the receipt of an application for the enlargement or extension of a manufactured home [land-lease development] in existence on April 22, 1987, the Commission may modify the requirements of this section, provided that to do so will not result in an overall improvement in the design or standards of the existing [land-lease development] .

[West Sacramento, California, Zoning Ordinance, 1995]

XII. PERMITTING AND SITE PLAN REVIEW AND PROCEDURE

Purpose. *Describes the proposal review procedure and identifies the information required to review an application for a manufactured home development.*

Notes. *In addition to providing information about the basic site plan submission and review requirements, these provisions may also include design standards that local officials believe will enhance further the design of new manufactured home developments. Poplar Bluff, Missouri, for example, establishes in its review requirements that "a site development plan* *which conforms to and preserves terrain, trees, shrubs, and rock formation is highly preferred." This is one of several general design standards included in the city's proposal review provisions. Albany, Oregon, requires that the applicant certify that a professional design team was used "in the design and development of the project."*

Site Plans

1. **Site plan review by planning board.** Prior to approval of a zoning permit by the zoning officer for the construction of a new or expansion of an existing manufactured housing development, a site plan shall be reviewed by the planning board. Said site plan shall meet the site plan requirements found in Article VII.

2. **Conditions.** In approving the site plans for manufactured housing developments, the planning board shall determine that adequate provision is made for the following:

    (a) Vehicular traffic to and from the development, and traffic internal to the development, including adequate access for emergency vehicle and personnel, postal service, and other public and private services and individuals who would require access to the premises.

    (b) Pedestrian traffic to and from the proposed manufactured home sites, common facilities, and parking areas on the premises.

    (c) Adequate types of common recreation areas, including any needed screening or landscaping.

    (3) **Final development plan.** Prior to the issuance of a certificate of occupancy, a final development plan indicating each manufactured home space and prepared in conformance with the subdivision regulations. . .shall be approved by the planning staff and recorded in the Office of the Register of Deeds. In addition, the corners of all manufactured home spaces shall be clearly marked on the ground with iron stakes.

    [Winston–Salem/Forsyth County, North Carolina,
    Unified Development Code, Section 2–5.43.K.]

Section 19–11. Design Review Standards

(a) **Purpose.** The purpose of the following standards is to provide the planning and zoning commission and the city council with standards to guide their review of a proposed development plan for a new manufactured home park or subdivision.

(b) **General design standards.** The development plan for a [land-lease development] or subdivision shall be studied to determine whether the proposed site development plan meets the requirements of this chapter and the following:

    (1) Each site development plan should be laid out and developed as a complete unit, in accordance with an integrated overall design.

    (2) The location for buildings, parking areas, walks, lighting, and appurtenant facilities should be arranged to be compatible with the surrounding land uses and any part of the site not used for buildings or other structures, or for parking or for roads or accessways, or recreation purposes, should be landscaped with grass, trees and shrubs.

    (3) There should be sufficient private open space in each lot to ensure that each resident has an adequate amount of light, air, and privacy.

    (4) Parking areas should be arranged to be safe and convenient from all directions within the development.

    (5) The street pattern should serve, not shape the lots in the development. Minor streets should feed at well spaced intervals into collector arterial streets.

    (6) A site development plan which conforms to and preserves terrain, trees, shrubs and rock formation is highly preferred.

    (7) Additional requirements as to landscaping, lighting, screening, accessways, building setbacks, and other site standards may be imposed by the city for

the protection of all property. Sanitary conditions as influenced by surface drainage and soil should be considered so as to prevent the accumulation of water upon the site.

    (8) The planning and zoning commission should determine that the proposed site development plan is consistent with good planning practice, consistent with the city's comprehensive plan, can be operated in a manner that is not detrimental to permitted uses in the surrounding area and is designed to promote the general welfare of the city.

(c) Development plan. A development plan shall be submitted at the time application is made for subdivision and/or conditional use approval. The following information is to be provided in the development plan:

    (1) Tract boundaries.

    (2) North point.

    (3) Topographical map of physical features.

    (4) Proposed streets, sanitary sewers, storm water sewer, water lines, and sidewalks in accordance with Appendix A, Subdivisions, Code of Ordinances, City of Poplar Bluff.

    (5) Location of manufactured home stands.

    (6) Type and location of protective screening.

    (7) Off-street parking spaces.

    (8) Common open space and improvements.

    (9) Accessory buildings.

    (10) Grading plan.

Section 19–12. Permits. Every manufactured home shall be inspected by the building inspector to insure conformance with the regulations of this chapter. Prior to the construction or installation of any manufactured home, a bundling permit must be obtained through the inspection department for a fee of $20.00.

[Poplar Bluff, Missouri, Code of Ordinances, 1982]

Section 36–27. Procedure

1. An application for an amendment for the Manufactured Home Zone to the zoning map shall follow all procedural requirements for amendments as set forth herein, and in addition shall include the following information:

    a. Site plan showing precise number, locations, and dimensions of all manufactured home lots, public or private drives or streets, illumination facilities, recreation or green areas, utilities, etc. Such site plans, if approved, shall form the basis for the issuance of a manufactured home park permit or as a preliminary subdivision study, whichever is the intention of the owner.

    b. Data as may be requested by the chief building official to determine that the proposed manufactured home development will comply with all legal requirements.

[Grand Island and Hall County, Nebraska,
Zoning Ordinance, Section 36–27]

Application Requirements

10.420. Professional Design Team. The applicant for proposed manufactured home [land-lease development] shall certify in writing that a registered architect or professional designer, a landscape architect, and a registered engineer or land surveyor licensed by the State of Oregon have been utilized in the design and development of the project.

10.430. Plot Plans Required. The application for a new or expansion of an existing manufactured home [land-lease development] shall be accompanied by 10 copies of the plot plan of the proposed [land-lease development]. The plot plan should show the general layout of the entire manufactured home [land-lease development] and should be drawn to a scale not smaller than 1" = 40'. In addition to the requirements of Section 8.120, the plan must include the following information:

(1) The location of adjacent streets and all private right-of-way existing and proposed within 300 feet of the development site.

(2) A legal survey.

(3) The boundaries and dimensions of the manufactured home [land-lease development].

(4) The location, dimensions, and number of each manufactured home space.

(5) The name and address of manufactured home [land-lease development].

(6) The scale and north point of the plan.

(7) The location and dimensions of each existing or proposed structure, together with the usage and approximate location of all entrances, height, and gross floor area.

(8) The location and width of accessways and walkways.

(9) The extent, location, arrangement, and proposed improvements of all off-street parking and loading facilities, open space, landscaping, fences and walls, and garbage receptacles.

(10) Architectural drawings and sketches demonstrating the planning and character of the proposed development.

(11) The total number of manufactured spaces.

(12) The location of each lighting fixture for lighting manufactured home spaces and grounds.

(13) The location of recreation areas, buildings, and area of recreation space in square feet.

(14) The point where the manufactured home [land-lease development] water and sewer system connects with the public system.

(15) The location of available fire and irrigation hydrants.

(16) An enlarged plot plan of a typical manufactured home space, showing location of the foundation, patio, storage space, parking, sidewalk, utility connections, and landscaping.

[Albany, Oregon, Development Code, 1994]

Information Required for Review of Residential Design Manufactured Homes (RDMH). The following shall be provided to the Pinellas County Zoning Division with all requests for review of RDMH per Section 506.13, Pinellas County Zoning Ordinance:

1. Recent actual photographs of all sides of the RDMH—photographs must depict the identical model to be used.

2. Exterior dimensions of the RDMH.

3. Type of roof materials to be used.

4. Pitch of roof and dimensions of the roof overhang.

5. Description of the exterior finish—photograph shall depict finish to be used.

6. Foundation plan.

7. Plot plan showing the placement of the RDMH on the lot.

Procedures for approval. Approval of residential design manufactured housing (RDMH) shall be authorized by the county administrator or his designee.

(1) Applications for approval of manufactured homes as RDMH shall be submitted to the county administrator or his designee in such form as may reasonably be required to make determinations. In particular, in addition to such information as is generally required for permits and as is necessary for administrative purposes, such applications shall include all information necessary to make determinations

as to conformity with the standards in this section, including photographs of all sides of the RDMH, exterior dimensions, roof pitch, roof materials, exterior finish, and other information necessary to make determinations.

(2) Actions by the county administrator or his designee; time limitations on determinations. Within seven days of receipt of the application and all required supporting materials, the county administrator or his designee shall make the determination as to conformity with the standards in this section and shall notify the applicant of the approval, conditional approval, or denial of the application. Conditional approval shall be granted only where the conditions and reasons therefor are stated in writing and agreed to by the applicant, and such conditions shall be binding upon the applicant. In the case of disapproval, the reasons therefor shall be stated in writing.

Standards for determination of similarity in exterior appearance. The following standards shall be used in determinations of similarity in appearance between residential design manufactured homes (RDMH), with foundations approved as provided in this subsection, and compatible in appearance with site-built housing which has been constructed in adjacent or nearby locations.

(1) **Minimum width of main body.** Minimum width of the main body of the RDMH as located on the site shall not be less than 20 feet, as measured across the narrowest portion. This is not intended to prohibit the offsetting of portions of the home.

(2) **Minimum roof pitch; minimum roof overhang; roofing materials.** Minimum pitch of the main roof shall be not less than three feet of rise for each 12 feet of horizontal run and minimum roof overhang shall be one foot. In cases where site-built housing generally has been constructed in adjacent or nearby locations with lesser roof pitches and/or roof overhangs of less than one foot, then the RDMH may have less roof pitch and overhang, similar to the site-built houses. In general, any roofing material other than a built-up composition roof may be used which is generally used for site-built houses in adjacent or nearby locations.

(3) **Exterior finish; light reflection.** Only material for exterior finish which is generally acceptable for site-built housing which has been constructed in adjacent or nearby locations may be used, provided, however, that reflection for such exterior shall not be greater than from siding coated with clean white gloss exterior enamel.

(4) **Approved foundations required in residential districts.** No RDMH shall be placed or occupied for residential use on a site in a residential district until such foundation plans have been submitted to and approved by the county administrator or his designee as to the appearance and durability of the proposed foundation and being acceptably similar or compatible in appearance to foundations of residences built on adjacent or nearby sites. All homes shall be placed on permanent foundations.

(5) **Site orientation of the manufactured home.** RDMHs shall be placed on lots in such a manner as to be compatible with and reasonably similar in orientation to the site-built housing which has been constructed in adjacent or nearby locations.

(6) **Garages, carports in residential neighborhoods where adjacent to nearby site-built homes which include garages and/or carports.** A RDMH shall be required to be provided with a garage and/or carport compatible with the RDMH and the site-built garages and/or carports constructed in adjacent or nearby locations.

(7) **Compatibility with nearby site-built housing.** RDMHs shall be compared to site-built housing in the neighborhood within the same zoning district. Approval for a RDMH shall not be granted unless it is found that the RDMH is substantially similar in size, siding, material, roof pitch, roof material, foundation and general appearance to site-built housing which may be permitted by the zoning and/or building code in the neighborhood in the same zoning district.

[Pinellas County, Florida, Zoning Ordinance, Section 138–1343, 1992]

Purpose. *To identify and define key words used in the ordinance.*

Notes. *Many communities continue to include outdated terms and definitions in their zoning ordinances. Some fail to make the distinction between manufactured homes (factory-built units built in compliance with the HUD Code) and mobile homes (factory-built units built prior to enactment of the HUD Code). Another term that needs to be updated in ordinances is "park." It is better to use the term "land-lease development" or "community." The term "park" is not an accurate description for a residential land-use development and is still strongly associated with mobile homes rather than manufactured housing. The term should be retained only in those communities that have pre-HUD Code mobile home or trailer parks.*

Defining Manufactured Housing as a Dwelling Unit

Dwelling–single (one) family. A permanent structure placed on a permanent foundation, having one or more rooms with provisions for living, sanitary, and sleeping facilities arranged for use of one or more individuals of the same family. The structure shall be located on a private lot and surrounded on all sides by a private yard. These dwellings shall include site-built, manufactured, and modular homes.

[Winnebago County, Wisconsin]

A General Category for Factory-built Homes

Factory-built housing. A structure designed for long-term residential use. For the purpose of these regulations factory-built housing consists of three types: modular, mobile homes, and manufactured homes.

[Huntington, West Virginia]

Defining Specific Factory-built Homes

Manufactured home. A dwelling unit fabricated in an off-site manufacturing facility for installation or assembly at the building site, bearing a label certifying that it is built in compliance with the Federal Manufactured Housing Construction and Safety Standards Act of 1974 (42 USC 5401, et. seq.), which became effective June 15, 1976.

[Huntington, West Virginia]

Mobile home. A transportable, factory-built home, designed to be used as a year-round residential dwelling and built prior to enactment of the Federal Manufactured Home Construction and Safety Standards Act of 1974 which became effective June 15, 1976 (42 U.S.C. 5403)   [Riley County, Kansas]

Modular home. A structure intended for residential use and manufactured off-site in accord with the [local or state] BOCA Basic Building Code.

[Montgomery County, Maryland]

Modular home. A structure constructed in a factory in accordance with the Uniform Building Code and bearing the appropriate insignia (gold) indicating such compliance. This definition includes "pre-fabricated," "panelized" and "factory-built" units.

[Kitsap County, Washington]

Defining Manufactured Home Developments

Manufactured home development. A general category of development that includes manufactured home subdivisions and manufactured home [land-lease developments].

Manufactured home subdivisions. A subdivision designed and/or intended for the sale of lots for siting manufactured homes.

Manufactured home subdivisions. Means a subdivision that is plotted for development as individual home sites for manufactured homes, modular homes, residential-design manufactured homes and site-built single-family dwellings to be placed on permanent foundations.

Manufactured housing community. Any piece of real property under single ownership or control for which the primary purpose is the placement of two or more manufactured homes for permanent residential dwellings and for the production of income. "Manufactured housing community" does not include real property used for the display and sale of manufactured homes, nor does it include real property used for seasonal recreational purposes only, as opposed to year-round occupancy.

[Issaquah, Washington]

Part 2. Springfield, Missouri, Zoning Ordinance Provisions For Manufactured Home Developments

Section 4–1500. R–MHC –Manufactured Home Community District

4–1501. Purpose. The R–MHC Manufactured Home Community District is established for manufactured homes in manufactured housing communities, which include manufactured housing developments and manufactured housing subdivisions, at low residential densities of approximately eight units per acre. It is intended that such manufactured housing communities shall be so located, designed, and improved as to provide:

A. a desirable residential environment;

B. protection from potentially adverse neighboring influences;

C. protection for adjacent residential properties;

D. principal access for vehicular traffic to collector or higher classification streets; and

E. accessibility equivalent to that for other forms of permitted residential development to public facilities, places of employment, and facilities for meeting commercial and service needs not met within the manufactured housing community. Other residential and supporting uses may also be permitted in such district.

4–1502. Permitted Uses in the R–MHC District

A. Single-family-detached dwellings, including [manufactured] homes, on individual building lots.

B. Single-family, semi-detached dwellings, in accordance with Section 3–3100, Cluster Subdivisions.

C. Manufactured housing developments.

D. Accessory apartments in owner-occupied, single-family-detached dwellings, in accordance with Section 5–2400.

E. Accessory Uses, as permitted by Section 5–1000.

F. Family day care homes, in accordance with Chapter 36, Article XI, Springfield City Code.

G. Group homes, residential, in accordance with Section 5–2500.

H. Home occupation uses, as permitted by Section 5–1100.

I. Outdoor storage areas in manufactured home developments, including storage areas for recreational vehicles. Such areas shall be screened from adjoining uses and shall occupy, in total, not more than 5 percent of the area of the manufactured home development. Use of such storage area shall be limited to the occupants of the manufactured home development.

J. Police and fire stations.

K. Schools, elementary and secondary, and schools or development centers for elementary- and secondary-school-age children with handicaps or development disabilities, on a minimum of five acres of land.

L. Temporary uses, as permitted by Section 5–1200.

M. Zero-lot-line construction, in accordance with Section 3–3200.

4–1503. Conditional Uses. The following conditional uses may be permitted provided they meet the provisions of, and a Conditional Use Permit is issued pursuant to, Section 3–3300 of this Article.

A. Adaptive use of nonresidential structures in accordance with Subsection 3–3310.B.5.

B. Cemeteries on a minimum of 10 acres of land.

C. Churches and other places of worship, including parish houses and Sunday schools, but excluding overnight shelters and temporary outdoor revivals, on a minimum of two acres of land, to provide sufficient land area for off-street parking, bufferyards, and proper site design to lessen impact on adjoining residential neighborhoods.

D. Clubhouses associated with any permitted use.

E. Noncommercial, not-for-profit neighborhood facilities, including indoor and outdoor recreational facilities, community centers, offices of property owners associations and maintenance facilities operated by a neighborhood or community organization or a property owners association in accordance with the provisions of Subsection 3–3310.B–1.

F. Public museums and libraries on a minimum of two acres of land.

G. Public service and public facility uses, as follow:
   1. Telecommunication towers and related facilities existing at the time the district is mapped, in accordance with Subsection 3–3310.B.2; and
   2. Water reservoirs, water standpipes, and elevated and ground-level water storage tanks.

4–1504. Use Limitations

A. In manufactured housing communities, recreational vehicles shall not be occupied as dwellings and manufactured home sales lots shall not be permitted.

B. All uses shall operate in accordance with the noise standards contained in Section 6–1500.

C. No use shall emit an odor that creates a nuisance as determined by Chapter 2A, Article X, Springfield City Code.

4–1505. Minimum Area Requirements for R–MHC Districts and for Individual Manufactured Home Communities; Minimum Number of Lots or Spaces to be Available in Manufactured Housing Development at Time of Opening. Where the district is established, the minimum area shall be 10 acres. In a manufactured housing development, the minimum number of lots or spaces completed and ready for occupancy before first occupancy is permitted shall be 30.

For manufactured housing developments, the tract shall comprise a single plot except where the site is divided by public streets or alleys or where the total property includes separate parcels for necessary utility plants, maintenance or storage facilities, or the like, with appropriate access from the manufactured housing development, provided that all lands involved shall be so dimensioned as to facilitate efficient design and management.

4–1506. Lot Size Requirements on Individual Lots
A. Minimum lot area: 5,000 square feet.

B. Minimum lot width: 45 feet.

C. Minimum lot depth: 75 feet.

4–1507. Maximum Density and Lot Size Requirements in Manufactured Housing Development. Maximum density in any manufactured housing development shall not exceed eight units per gross acre provided the requirements of Subsection 1–1322 are met. For purposes of these regulations, gross acreage is to be computed as all area within the exterior boundaries of the property, including streets, common open space, lands occupied by management offices and community buildings, lands occupied by manufactured home lots, and lands occupied by utilities installations. Lots for placement of manufactured homes in manufactured housing developments shall be at least 4,000 square feet in area with no dimension less than 40 feet. The limits of each manufactured home lot shall be shown on the site plan and shall be clearly marked on the ground by permanent flush stakes, markers, or other suitable means.

4–1508. Bulk and Open Space Regulations on Individual Lots

A. Maximum structure height:
   1. When side yards do not exceed 15 feet in width: 35 feet or 2.5 stories above the finished grade.
   2. When side yards exceed 15 feet in width: 3 stories above the finished grade.
   3. Accessory structures: 16 feet except storage buildings which shall not exceed 10 feet.

B. Minimum yard requirements (additional bufferyard may be required by Subsection 4–1411):
   1. Front yard: 25 feet, or as required by Section 5–1300.
   2. Side yards: 5 feet, or as required by Section 5–1300.
   3. Rear yard: 20 percent of the lot depth, but not less than 10 feet nor more than 25 feet be required.
   4. However, in no event may a structure be erected closer to the center line of an existing or planned street than as prescribed below, except as permitted by Subsection 1–1317.B.

Street Classification	Required Setback from Right-of-Way Center Line
Freeway	150 feet plus the required yard setback
Expressway	65 feet plus the required yard setback
Primary Arterial	50 feet plus the required yardsetback
Secondary Arterial	35 feet plus the required yard setback
Major Collector	30 feet plus the required yard setback
Residential Collector	25 feet plus the required yard setback
Commercial/Industrial Local	30 feet plus the required yard setback
Residential Local	25 feet plus the required yard setback
Highway Access Road	20 feet plus the required yard setback

C. **Maximum building coverage (including accessory buildings):** 40 percent.

D. Minimum open space. Not less than 30 percent of the total lot area shall be devoted to open space including required yards and bufferyards unless modified in accordance with Subsection 6–1215. Open space shall not include areas covered by buildings, structures, parking areas, driveways, and internal streets. Open space shall contain living ground cover and other landscaping materials.

E. **Maximum impervious area.** The combined area occupied by all main and accessory buildings or structures, parking areas, driveways, and any other surfaces which reduce and prevent absorption of stormwater shall not exceed 70 percent of the total lot area unless modified in accordance with Subsection 6–1215.

4–1509. Bulk and Open Space Requirements in Manufactured Housing Developments

A. Maximum structure height: 35 feet or 2.5 stories above the finished grade.

B. Minimum yard along exterior boundaries (additional bufferyard may be required by Subsection 4–1411): 25 feet. Where a manufactured housing development adjoins a public street or a residential district, including a R–MHC district, without an alley or other permanent open space at least 25 feet in width, the required yard shall not contain garages, carports, recreational shelters, storage structures, or any other structure generally prohibited in yards adjacent to streets or in residential districts. No direct vehicular access to individual lots shall be permitted through such yards and no group parking facilities or active recreation areas shall be allowed therein. Where the adjoining district is nonresidential, such yards may be used for group or individual parking; active recreation facilities; or carports, recreational shelters, or storage structures.

C. However, in no event may a structure be erected closer to the center line of an existing or planned street than as prescribed below:

Street Classification	Required Setback from Right-of-Way Center Line
Freeway	150 feet plus the required yard setback
Expressway	65 feet plus the required yard setback
Primary Arterial	50 feet plus the required yard setback
Secondary Arterial	35 feet plus the required yard setback
Major Collector	30 feet plus the required yard setback
Residential Collector	25 feet plus the required yard setback
Commercial/Industrial Local	30 feet plus the required yard setback
Residential Local	25 feet plus the required yard setback
Highway Access Road	20 feet plus the required yard setback

D. **Minimum open space**. Not less than 20 percent of the total area of the manufactured housing development shall be devoted to open space including required yards and bufferyards unless modified in accordance with Subsection 6–1215. Open space shall not include areas covered by buildings, structures, parking areas, driveways, and internal streets. Open space shall contain living ground cover and other landscaping materials.

E. **Maximum impervious area.** The combined area occupied by all main and accessory buildings or structures, parking areas, driveways, and any other surfaces which reduce and prevent absorption of water shall not exceed 80 percent of the total area of the manufactured housing development unless modified in accordance with Subsection 6–1215.

4–1510. Density Requirements. Reserved.

4–1511. Design Requirements on Individual Lots

A. A site plan meeting the requirements of Section 3–3000 shall be submitted and approved for all uses except single-family-detached dwellings and manufactured homes on individual lots.

B. A plot plan meeting the requirements of Subsection 3–1103 shall be submitted and approved for all single-family-detached dwellings and manufactured homes on individual lots.

C. A landscaping plan meeting the requirements of Section 6–1200 and 6–1300 shall be submitted and approved for all uses except single-family-detached dwellings and manufactured homes on individual lots.

D. All off-street parking lots and vehicular use areas for permitted nonresidential uses shall be screened from all residential uses in accordance with Section 6–1200.

E. Refuse storage areas for permitted nonresidential uses shall be screened from view in accordance with Section 6–1000.

F. Required front yards shall be landscaped with grass, ground cover, plants, shrubs or trees. Decorative landscaping materials such as rock, bark, and mulch are also permitted. Impervious surfaces in required front yards shall be minimized and shall be limited to driveways leading to off-street parking areas located outside the required front yard and walkways necessary for access to structures on the property. Circular drives are permitted if sufficient frontage is available and if approved by the Traffic Engineer.

G. Storage of maintenance or other equipment incidental to any permitted or conditional use except a single-family-detached dwelling or manufactured home on an individual lot shall be screened from view in accordance with the provisions of Section 6–1000.

H. Skirting shall be placed around manufactured homes that are not placed on a permanent foundation. Such skirting shall be similar in appearance to materials used for permanent foundations or the siding of the manufactured home to which it is to be attached.

I. Mechanical and electrical equipment, including air conditioning units, shall be screened from view in accordance with Section 6–1000.

4–1512. Design Requirements in Manufactured Housing Developments

A. A site plan meeting the requirements of Section 3–3000 shall be submitted and approved.

B. A landscaping plan meeting the requirements of Section 6–1200 and 6–1300 shall be submitted and approved.

C. There shall not be less than 15 feet between manufactured homes or any other buildings located in a manufactured housing development, and location on the lot shall be suitable for the type of manufactured home proposed, considering size and manner of support, and any improvements necessary on the lot for the support or anchoring of the type of manufactured home proposed shall be provided to the manufactured home so supported and/or anchored before occupancy. Parking spaces for each manufactured home do not have to be provided on each lot, however one of the two parking spaces required shall be located within 100 feet of the lot served.

D. Storage of maintenance or other equipment incidental to any permitted or conditional use shall be screened from view in accordance with Section 6–1000.

E . Refuse storage areas shall be screened from view in accordance with Section 6–1000.

F . Lighting shall be designed to reflect away from any adjacent residential areas and in accordance with Section 6–1400.

G . Mechanical and electrical equipment, including air conditioning units, shall be screened from view in accordance with Section 6–1000.

H. Skirting shall be placed around manufactured homes that are not placed on a permanent foundation. Such skirting shall be similar in appearance to materials used for permanent foundations or the siding of the manufactured home to which it is to be attached.

4–1513. Bufferyard Requirements. Whenever any development in an R–MHC district is located adjacent to a different zoning district or a nonresidential use in an R–MHC district is located adjacent to a residential use in an R–MHC district, screening and a bufferyard shall be provided in accordance with Sections 6–1000 and 6–1200.

Rules of Interpretation and Definitions

Manufactured home. A factory-built structure which bears the seal of the State of Missouri public service commission, U.S. Department of Housing and Urban Development, or its agent, and . . . is equipped with the necessary service connections and made so as to be readily movable as a unit on its own running gear and designed to be used as a dwelling unit with or without a permanent foundation.*

Manufactured housing development. A site with required improvements and utilities for the long-term placement of manufactured homes for dwelling purposes. Services and facilities for residents of the development may also be included on the site.

Manufactured housing subdivision. A development containing lots intended primarily for the individual placement of manufactured homes for dwelling purposes.

Mobile home. A transportable, factory-built home, designed to be used as a year-round residential dwelling and built prior to the enactment of the Federal Manufactured Housing Construction and Safety Standards Act of 1974, which became effective June 16, 1976.

Mobile home park. *See* **Manufactured housing development.**

Modular home. A factory-built transportable structure which bears the seal of the State of Missouri public service commission or is built to the BOCA Basic Building Code as adopted by the City of Springfield and which does not have its own running gear and is designed to be used as a dwelling unit with a permanent foundation.

---

* **Note:** Language determined to be only applicable to local circumstances or a matter of local preference has been omitted from this definition.

Project Directory

Canyon View Estates
Santa Clarita, California

Developer:
Canyon View Partnership
20001 Canyon View Dr.
Santa Clarita, CA 91351
805-298-1000

Elmhurst Infill Development
Laurel Courts
Martin Luther King Jr. Plaza
(proposed)
Oakland, California

Developer:
Paul Wang and Associates
950 Regal Road
Berkeley, CA 94708
415-524-6752

Architect/Planner:
Paul Wang and Associates

Manufacturer:
Silvercrest Western Homes
   Corporation
299 North Smith Ave.
Corona, CA 91720
909-734-6610

Lexington
Apex, North Carolina

Developer:
Canterbury Communities, Inc.
(a subsidiary of Pulte Homes
   Corporation)
Box 1488
Cary, NC 27512
919-387-9531

Architect/Planner:
Jerry Turner & Associates
905 Jones Franklin Road
Raleigh, N.C. 27606

Lido Peninsula
Newport Beach, California
(Redevelopment Project)

Owner:
John L. Curci
710 Lido Park Dr.
Newport Beach, CA 92659
714-673-1060

Developer:
Bessire and Casenhiser, Inc.
661 Brea Canyon Road, Ste. 7
Walnut, CA 91789
909-594-0501

Designer (replacement homes):
RGC
20 Corporate Plaza
Newport Beach, CA 92660
714-720-9881

Manufacturer:
Silvercrest Western Homes
   Corporation
299 North Smith Ave.
Corona, CA 91720
909-734-6610

Manufactured Housing Institute Urban Design Project
(various sites listed below)

Project Coordinator:
Andrew Scholz
Manufactured Housing Institute
2101 Wilson Blvd.
Suite 610
Arlington, VA 22201-3062
703-558-0652

Architect/Planner:
Susan Maxman and Partners, Ltd.,
   Architects
123 S. 22nd St.
Philadelphia, PA 19103
215-977-8662

Birmingham, Alabama
Developer:
Smithfield, Inc.
20 North 13th Ave.
Birmingham, AL 35204
205-322-3117

Manufacturer:
Cavalier Homes of Alabama
P.O. Box 300
Addison, AL 35540
205-747-1575

Louisville, Kentucky
Developer:
Neighborhood Development
    Corporation
1244 S. 4th St.
Louisville, KY 40205
502-637-2591

Manufacturer:
New Era Building Systems, Inc.
P.O. Box 269
Strattanville, PA 16258
814-764-5581

Milwaukee, Wisconsin
Developer:
Community Development
    Corporation of Wisconsin
152 W. Wisconsin Ave., Ste. 314
Milwaukee, WI 53203
414-277-9630

Manufacturer:
Schult Homes Corporation
P.O. Box 151
Middlebury, IN 46440
219-825-6210

Washington, D.C.
Developer:
Marshall Heights Community
    Development Organization, Inc.
3917 Minnesota Ave., NE, 2nd floor
Washington, D.C. 22219
202-396-1200

Manufacturer:
Schult Homes Corporation
P.O. Box 151
Middlebury, IN 46440
219-825-6210

Wilkinsburg, Pennsylvania
Developer:
Action Housing Inc.
2 Gateway Center, 18th Floor
603 Stanwick St.
Pittsburgh, PA 15222

Manufacturer:
New Era Building Systems
P.O. Box 269
Strattanville, PA 16258
814-764-5581

NextGen
(research project on the future of
    manufactured housing)
Steven Winter Associates, Inc.
Building Systems Consultants
50 Washington St.
Norwalk, CT 06854
203-857-0200

New Colony Village
Baltimore, Maryland
Developer:
Corridor 1 LP
P.O. Box 39
Columbia, MD 21045
410-792-7770

Architect:
Hackworth Architecture
1927 Post Alley
Seattle, WA 98100
206-443-1181

Manufacturer:
Schult Homes Corporation
P.O. Box 151
Middlebury, IN 46440
219-825-6210

Noji Gardens
Seattle, Washington
Developer:
HomeSight
3405 South Alaska St.
Seattle, WA 98118
206-723-7924

Architect:
John T. McLaren
2727 Fairview Ave. East, Ste. A
Seattle, WA 98102
206-325-9890

Manufacturer:
Marlette Homes, Inc.
P.O. Box 910
Hermiston, OR 97838
514-567-5546

Rancho Viejo
Escondido, California
Developer:
Robert L. Childers Co., Inc.
4802 Lake Park Place
Escondido, CA 92026
619-576-8000

Riverbrook (Community Upgrade)
New Haven, Michigan
Architect/Planner:
Donald Westphal, ASLA
512 Madison Ave.
Rochester, MI 48307
248-651-5518

Village of Rosa Vista
Mesa, Arizona

Developer:
Homefree Village Resorts
1400 S. Colorado Blvd., Ste. 410
Denver, CO 80222
305-757-3002

Architect/Planner:
Andres Duany and Elizabeth
    Plater-Zyberk
Architects and Town Planners
1023 Southwest 25th Ave.
Miami, FL 33135
305-644-1023

Manufacturer:
Cavco
2502 West Durango
Phoenix, AZ 85009
602-270-3554

Santiago Estates
Sylmer, California
Developer:
Santiago Estates
13691 Gavina Ave.
Sylmer, CA 91342
818-362-5022

Shelby West
Shelby Township, Michigan

Developer:
Shelby Forest Associates
12741 S. Saginaw St.
Grand Blanc, MI 48439

Architect/Planner:
Donald Westphal, ASLA
512 Madison Ave.
Rochester, MI 48307
248-651-5518

Adopted by the Chapter Delegate Assembly on September 20, 1997
Ratified by the Board of Directors on September 21, 1997

Statement of the Issues

An important element of APA's housing policy is the promotion of decent housing affordable to low- and moderate-income households in suitable living environments. Specifically, the policy urges all levels of government to recognize manufactured homes as an acceptable form of housing and a viable alternative to more costly site-built housing. This is particularly important as housing subsidies at the federal level are reduced. This policy is not intended to address siting issues with respect to mobile home parks but is a guide for the standards to be employed in the siting and design of manufactured homes on individual lots.

Findings

American Planning Association Policy Guide on Manufactured Housing

Since 1976, the U.S. Department of Housing and Urban Development (HUD) has regulated manufactured homes under the Manufactured Home Construction and Safety Standards, which are commonly referred to as the HUD Code. At that time, these housing units were called "mobile homes," but, in 1980, this designation was changed to "manufactured home" in recognition of the more durable and less mobile nature of these homes. Once sited, these homes are rarely moved.

About one-third of all residential units built in this country in 1995 were factory-built. Manufactured homes (HUD Code) that consist of one or more finished modules constructed on a permanent chassis used to transport the unit to its site accounted for 68 percent of those factory-built homes. Panelized homes, constructed of factory-made panels assembled on site, accounted for 26 percent. The remaining 6 percent of factory-built homes were modular units, built with one or more three-dimensional components and transported to the site on a flat-bed truck.

A 1985 nationwide survey by APA revealed that manufactured homes have not only become safer and more durable since enactment of the HUD Code in 1976, but their appearance has improved significantly. At the same time, public acceptance of manufactured homes has increased, and some communities have revised their zoning and subdivision standards that govern manufactured homes and now permit such homes by right even in their most restrictive single-family districts.

It also is important that there is a balance between housing needs and the need for stability of existing neighborhoods. Manufactured housing, if not properly placed and sited, can conflict with established neighborhood development patterns. Owners will pay some price for setting a home in an urban environment. This may include additional costs for underpinning, roof material, siding, and other components. These are factors that can be, and are, applied equally to site-built homes. The inherent resistance to manufactured housing, however, may require more diligence in the design and administration of regulations.

A growing number of communities are allowing manufactured homes on vacant infill lots in built-up neighborhoods. At the same time, many developers who have never worked with manufactured homes before are taking advantage of greater public acceptance of this type of housing and are using it in new subdivision development. This may create interesting opportunities for manufactured housing as well as existing neighborhoods. Well-conceived pilot and demonstration projects may be used as a means for erasing stereotypes and encouraging more widespread use of manufactured homes in urban settings.

Local public housing authorities and private not-for-profit organizations are exploring the use of manufactured homes to provide affordable housing to low- and moderate-income families. In recent years, there has been increasing interest in locating manufactured home production plants in inner-city areas where both jobs and affordable housing are badly needed.

Many communities, however, continue to carry over outmoded stereotypes and exclude all types of manufactured homes from residential neighborhoods, even though the designs may be very similar in appearance and size to site-built housing.

Policy Positions

General Policy Position. APA National and Chapters support the use of manufactured homes on individual lots in compliance with the HUD Code as an acceptable form of affordable housing.

Reasons to support general policy position:

a. The manufactured home is a major source of housing for young families, first-time home buyers, older adults, and others with limited income. In 1995, one-third of all new single-family homes sold were manufactured homes. Support for this motion would be in keeping with APA's social equity policies.

b. HUD has made expanded use of manufactured housing for affordable homeownership a key element of its National Homeownership Strategy. APA is one of a number of national associations that make up HUD's National Partners in Homeownership and, along with other members of the partnership, is helping to formulate and carry-out this strategy.

c. Support for manufactured homes and other forms of lower-cost housing is consistent with APA's existing housing and social equity policies, which promote decent housing affordable to low- and moderate-income households in suitable living environments.

Specific Policy Position 1. APA National and Chapters support reasonable, cost-effective standards, including design standards, and administrative procedures for manufactured homes in local zoning ordinances and regulations.

Reasons to support specific policy position 1:

a. Affordability is the single most important attribute of manufactured homes. Land-use regulations should not diminish the cost savings this form of housing offers. Local zoning and subdivision standards can be crafted so that they encourage good siting and design without unnecessarily limiting the use of manufactured housing. These standards and the process for applying them should be no more restrictive for manufactured housing units than for conventionally constructed housing units. At the same time, it must be recognized that there may be some costs incurred to make manufactured housing compatible with existing neighborhoods in an infill environment, thereby resulting in development that may be more costly than typical subdivisions.

b. The HUD Code is a preemptive, uniform construction code that ensures that a manufactured home, regardless of where it is built in the U.S., will meet certain publicly adopted standards related to health, safety, and welfare.

c. A growing number of states have enacted laws that prohibit the exclusion and unfair regulatory treatment of manufactured homes. Some states, moreover, call for parity in the regulation of manufactured homes and site-built housing.

d. Manufactured homes should be allowed as a permitted use in residential zoning districts at the permitted density for the district.

Specific Policy Position 2. APA National and Chapters should develop model definitions and siting standards.

Reasons to support specific policy position 2: APA can play a leadership role in standardizing regulations to be implemented by communities that want to ensure more equitable treatment of manufactured housing and to create affordable housing opportunities for their citizens.

Specific Policy Position 3. APA National and Chapters encourage states to take steps to ensure that installation and anchoring requirements for manufactured homes are adequate. Where unique environmental conditions exist, specific life safety standards should be coordinated with those required by HUD. These standards should be consistent with those set forth in the American National Standards Institute's publication, *Manufactured Home Installation* (1994).

Reasons to support specific policy position 3: About half of the states have adopted installation standards for manufactured homes that require these homes to be installed on properly engineered foundation systems. When properly anchored, manufactured homes perform on an equal basis with site-built dwellings in unique environmental conditions.

Specific Policy Position 4. APA National and Chapters support periodic review and revision of the HUD Code to ensure that these standards adequately protect the public health, safety, and welfare.

Reasons to support specific policy position 4: Typically, model codes such as BOCA (Basic Building Code) and ICBO (Uniform Building Code) are revised on a three-year cycle. It makes sense for the HUD Code to undergo a similar revision schedule.

Exceptions. Exceptions from the General Policy Position or Specific Policy Positions supported by specific findings and reasoning: *None to date*

Amendments. This policy is subject to amendment for the purposes of the following:
1. adding findings or supplementing previous findings with new data or interpretations; and

2. adding Specific Policy Positions based on new findings or reasoning that tend to add to or qualify, but not reject entirely, the General Policy Position, one or more Specific Policy Positions, or one or more Exceptions from Policy Positions.

There have been no amendments made to this policy as of August 1, 1998.

Bibliography

American Planning Association. 1994. *Planning and Community Equity*. Chicago: APA.

Sanders, Welford. 1993. *Manufactured Housing Site Development Guide*. Planning Advisory Service Report No. 445. Chicago: American Planning Association, April.

_____. 1986. *Regulating Manufactured Housing*. Planning Advisory Service Report No. 398. Chicago: American Planning Association, April.

Manufactured Housing State Associations (through July 1998)

Alabama

Alabama Manufactured. Housing
   Institute
Sherry Norris, Executive Director
4274 Lomac St.
Montgomery, AL 36106
(334) 244-7828
(334) 244-9339 (fax)

Arizona

Manufactured Housing Industry of
   Arizona
Gub Mix, Executive Director
P.O. Box 24049
Tempe, AZ 85285-4049

Physical Location:
4525 S. Lakeshore Dr., Ste. 5
Tempe, AZ 85282
(602) 456-6530
(602) 456-6529 (fax)

Arkansas

Arkansas Manufactured Housing
   Association
J. D. Harper, Executive Director
2500 McCain Place, Ste. 203
N. Little Rock, AR 72116
(501) 771-0444
(501) 771-0445 (fax)
arkmha@aol.com

California

California Manufactured Housing
   Institute
Bob West, Executive Director
10630 Town Center Dr., Ste. 120
Rancho Cucamonga, CA 91730
(909) 987-2599
(909) 989-0434 (fax)
info@cmhi.org

WMA
Sheila Dey, Executive Director
1007 7th St., Ste. 300
Sacramento, CA 95814
(916) 448-7002
(916) 448-7085 (fax)

Colorado

Colorado Manufactured Housing
   Association
Bonnie Geiger, Executive Director
1410 Grant St., Ste. D-110
Denver, CO 80203
(303) 832-2022
(303) 830-0826 (fax)
ghstar@ix.netcom.com

Connecticut

Connecticut Manufactured Housing
   Association
Joe Mike, Executive Director
P.O. Box 605
Bristol, CT 06011
(860) 278-7650
(860) 278-7671 (fax)
atlantich@aol.com

Delaware

Delaware Manufactured Housing
   Association
Marcene Gory, Executive Director
9 East Lockerman
Treadway Towers, Ste. 309
Dover, DE 19901
(302) 678-2588
(302) 678-4767 (fax)

First State Manufactured Housing
   Association
Phyllis McKinley, Executive Director
P.O. Box 1829
Dover, DE 19903
(302) 674-5868
(302) 674-5960 (fax)
fsmhi@magpage.com

Florida

Florida Manufactured Housing
   Association
Frank Williams, Executive Director
2958 Wellington Circle, Ste. 100
Tallahassee, FL 32308
(850) 907-9111
(850) 222-7957 (fax)
fmha@nettally.com

Georgia

Georgia Manufactured Housing
  Association
Charlotte Gattis, Executive Director
1000 Circle 75 Parkway, #060
Atlanta, GA 30339
(770) 955-4522
(770) 955-5575 (fax)
cagattis@mindspring.com

Idaho

Idaho Manufactured Housing
  Association
Gub Mix, Executive Director
P.O Box 201
Sun Valley, ID 83353
(208) 343-1722
(208) 622-6523 (fax)

Illinois

Illinois Manufactured Housing
  Association
Mike Marlowe, Executive Director
3888 Peoria Road
Springfield, IL 62702
(217) 528-3423
(217) 544-4642 (fax)
earthlink.net

Illinois Housing Institute
Rob Patterson, Executive Director
8700 W. Bryn Mawr, Ste. 800S
Chicago, IL 60631
(773) 714-4920
(773) 714-4925 (fax)

Indiana

Indiana Manufactured Housing
  Association
Dennis Harney, Executive
  Vice President
3210 Rand Road
Indianapolis, IN 46241
(317) 247-6258
(317) 243-9174 (fax)
imharvic@ix.netcom.com

Iowa

Iowa Manufactured Housing
  Association
Joe Kelly, Executive Vice President
1400 Dean Ave.
Des Moines, IA 50316
(515) 265-1497
(515) 265-6480 (fax)
house@netins.net

Kansas

Kansas Manufactured Housing
  Association
Martha Neu Smith, Executive Director
214 S.W. 6th St., Ste. 206
Topeka, KS 66603-3719
(913) 357-5256
(913) 357-5257 (fax)
kmha@cjnetworks.com

Kentucky

Kentucky Manufactured Housing
  Institute
Thad Vann, Executive Director
2170 US 127 South
Frankfort, KY 40601
(502) 223-0490
(502) 223-7305 (fax)

Louisiana

Louisiana Manufactured Housing
  Association
Steve Duke, Executive Director
4847 Revere Ave.
Baton Rouge, LA 70808
(504) 925-9041
(504) 925-1208 (fax)
steve@lmha.com

Maine

Manufactured Housing Association
  of Maine
Robert Howe, Executive Director
3 Wade St., Liscomb Bldg.
P.O. Box 1990
Augusta, ME 04330-1990
(207) 623-2204
(207) 622-4437 (fax)
howe@maine.com

Maryland

Manufactured Housing Institute
  of Maryland, Inc.
Lowell Cochran, Executive Director
P.O. Box 1158
Hagerstown, MD 21740-1158
(301) 797-5341
(301) 797-6836 (fax)

Massachusetts

Massachusetts Manufactured Housing
  Association
Richard Norton, Executive Director
P.O. Box 5963
Marlborough, MA 01752
(508) 460-9523
(508) 460-9360 (fax)

Michigan

Michigan Manufactured Housing
    Association
Tim DeWitt, Executive Director
2222 Association Dr.
Okemos, MI 48864-5978
(517) 349-3300
(517) 349-3543 (fax)
 michhome@rust.net

Minnesota

Minnesota Manufactured Housing
    Association
Mark Brunner, Executive
    Vice President
555 Park St., Ste. 400
Saint Paul, MN 55103
(612) 222-6769
(612) 222-6913 (fax)
evpmail@mmmfghome.org

Mississippi

Mississippi Manufactured Housing
    Association
Jennifer Hanson, Executive Director
P.O. Box 54266
Pearl, MS 39288-4266
(601) 939-8820
(601) 939-7988 (fax)

Physical Location:
1001 Airport Road
Flowood, MS 39208

Missouri

Missouri Manufactured Housing
    Association
Joyce Baker, Executive Director
P.O. Box 1365
4748 Country Club Dr.
Jefferson City, MO 65102
(573) 636-8660
(573) 636-4912 (fax)
mmha@mail.cscdata.net

Montana

Montana Manufactured Housing
    Association
Stuart Doggett, Executive Director
7 West 6th Ave., Ste. 507
Helena, MT 59601
(406) 442-2164
(406) 442-8018 (fax)
stuart@initco.net

Nebraska

Nebraska Manufactured Housing
    Institute, Inc.
Martin Huff, Executive Director
5300 West O Str.
Lincoln, NE 68528
(402) 475-3675
(402) 475-1359 (fax)

Nevada

Nevada Manufactured Housing
    Association
Gub Mix, Executive Director
P.O Box 201
Sun Valley, ID 83353
(702) 737-7778
(208) 622-6523 (fax)

New Hampshire

New Hampshire Manufactured
Housing Association
Robert Ruais, Executive Director
9 North Acres
Manchester, NH 03104
(603) 629-9369
(603) 627-1932 (fax)

New Jersey

New Jersey Manufactured Housing
    Association
Judith A. Thornton, Executive Director
P.O. Box 3172
Trenton, NJ 08619
(609) 588-9040
(609) 587-6697 (fax)

New Mexico

New Mexico Manufactured Housing
    Association
Mark Duran, Executive Director
6400 Uptown Blvd., Ste. 540 West
Albuquerque, NM 87110
(505) 830-3764
(505) 830-3772 (fax)
nmmha@ix.netcom.com

New York

New York Manufactured Housing
    Association
Nancy Geer, Executive Director
421 New Karner Road
Albany, NY 12205-3809
(518) 464-5087
(518) 464-5096 (fax)
nymha@frontier.net

North Carolina

North Carolina Manufactured
  Housing Institute
Steve Zamiara, Executive Director
P.O. Box 58648
Raleigh, NC 27658-8648
(919) 872-2740
(919) 872-4826 (fax)
ncmhi@aol.com

North Dakota

North Dakota Manufactured Housing
  Association
Executive Vice President
P.O. Box 2681
Bismark, ND 58502
(701) 667-2187
(701) 667-2187 (fax)

Ohio

Ohio Manufactured Housing Association
Tim Williams, Executive
  Vice President
201 Bradenton Ave. Ste. 100
Dublin, OH 43017
(614) 799-2340
(614) 799-0616 (fax)
williamst@ee.net

Oklahoma

Manufactured Housing Association
  of Oklahoma
Deanna Fields, Executive Director
6400 S. Shields
Oklahoma City, OK 73149
(405) 634-5050
(405) 634-5355 (fax)
mhao@mhao.com

Oregon

Oregon Manufactured Housing
  Association
Don Miner, Executive Director
2255 State St.
Salem, OR 97301
(503) 364-2470
(503) 371-7374 (fax)
omha@cyberhighway.net

Manufactured Housing Communities
  of Oregon, Inc.
Sample Lindholm, Executive Director
3857 Wolverine St., NE, Ste. 22
Salem, OR 97305
(800) 488-6426
(503) 391-4652 (fax)
sample@cyberis.net

Pennyslvania

Pennsylvania Manufactured Housing
  Association
Mary Gaiski, Executive Vice President
P.O. Box 248,
Routes 114 & I-83
New Cumberland, PA 17070
(717) 774-3440
(717) 774-5596 (fax)
mary@pmhainet.pmha.org

South Carolina

Manufactured Housing Institute
  of South Carolina
Mark Dillard, Executive Director
P.O. Box 1781
Columbia, SC 29202
(803) 771-9046
(803) 771-7023 (fax)
mhisc@scsn.net

South Dakota

South Dakota Manufactured Housing
  Association
Jerry Biedenfeld, Executive Director
P.O. Box 7077
Pierre, SD 57501

Physical Address:
203 Lindsey Trail
Ft. Pierre, SD 57532
(605) 224-2540
(605) 223-2061 (fax)
jerry@sb.cybernex.net

Tennessee

Tennessee Manufactured Housing
  Association
Bonita Hamm, Executive Director
240 Great Circle Road,Ste. 332
Nashville, TN 37228
(615) 256-4733
(615) 255-8869 (fax)

Texas

Texas Manufactured Housing
  Association
Will Ehrle, President and
  General Counsel
P.O. Box 14428
Austin, TX 78761
(512) 459-1221
(512) 459-1511 (fax)
texasmha@aol.com

Utah

Utah Manufactured Housing
    Association
Gub Mix, Executive Director
P.O. Box 201
Sun Valley, ID 83353
(702) 737-7778
(208) 622-6523 (fax)

Vermont

Vermont Manufactured Housing
    Association
Dave Atkins, Executive Director
165 Kellogg Road, Box 700
Colchester, VT 05446
(802) 879-7712
(802) 879-7710 (fax)

Virginia

Virginia Manufactured Housing
    Association
Ron Dunlap, Executive Director
8413 Patterson Ave.
Richmond, VA 23229
(804) 750-2500
(804) 741-3027 (fax)
vamha@i2020.net

Washington

Washington Manufactured Housing
    Association
Joan Brown, Executive Director
1530 Evergreen Park Dr., SW
Olympia, WA 98502
(360) 357-5650
(360) 357-5651 (fax)
whmha@nwpride.org

Manufactured Housing Communities
    of Washington, Inc.
Ken Spencer, Executive Director
509 E. 12th Ave., S.E., Ste. 7
Olympia, WA 98501
(360) 753-8730
(360) 753-8731 (fax)

West Virginia

West Virginia Manufactured Housing
    Association
Leff Moore, Executive Director
205 First Ave.
Nitro, WV 25143
(304) 727-7431
(304) 727-1172 (fax)
wvasoff@newwave.net

Wisconsin

Wisconsin Manufactured Housing
    Association
Ross Kinzler, Executive Director
202 State St., Ste. 200
Madison, WI 53703
(608) 255-3131
(608) 255-5595 (fax)
ross@wmha.org

Wyoming

Wyoming Manufactured Housing
    Association
Laurie Urbigkit, Executive Director
P.O. Box 1493
Riverton, WY 82401
(307) 857-6001
(307) 857-0537 (fax)

National Association

Manufactured Housing Institute
Andy Scholz, Vice President,
    Land-Use and Retail Activities
Eric Alexander, Manager,
    Land-Use Activities
2101 Wilson Blvd., Ste. 610
Arlington, VA 22201
(703) 558-0400
(703) 558-0401 (fax)
andy@mfghome.org
eric@mfghome.org